Random Processes
for Engineers
A PRIMER

Random Processes
for Engineers

A PRIMER

Arthur David Snider

CRC Press
Taylor & Francis Group
Boca Raton London New York

CRC Press is an imprint of the
Taylor & Francis Group, an **informa** business

MATLAB® is a trademark of The MathWorks, Inc. and is used with permission. The MathWorks does not warrant the accuracy of the text or exercises in this book. This book's use or discussion of MATLAB® software or related products does not constitute endorsement or sponsorship by The MathWorks of a particular pedagogical approach or particular use of the MATLAB® software.

CRC Press
Taylor & Francis Group
6000 Broken Sound Parkway NW, Suite 300
Boca Raton, FL 33487-2742

First issued in paperback 2020

© 2017 by Taylor & Francis Group, LLC
CRC Press is an imprint of Taylor & Francis Group, an Informa business

No claim to original U.S. Government works

ISBN-13: 978-1-4987-9903-4 (hbk)
ISBN-13: 978-0-367-65635-5 (pbk)

Visit the Taylor & Francis Web site at
http://www.taylorandfrancis.com

and the CRC Press Web site at
http://www.crcpress.com

Contents

Preface

There are a lot of authoritative, comprehensive, and axiomatically correct books on random processes, but they all suffer from lack of accessibility for engineering students starting out in the area. This book fills the gap between the undergraduate engineering statistics course and the rigorous approaches. A refresher on the prerequisite topics from basic statistics is given in the first chapter.

Some of the features that distinguish the book from other resources are the following:

1. "Probability spaces" based on measure theory and sigma-algebras are appropriate for addressing some sticky philosophical questions ("How can every possible outcome have probability zero while one of them certainly occurs?"), but this esoteric machinery is not necessary for solving practical problems. This book only discusses them sufficiently to introduce the issues and whet the readers' appetite for rigorous approaches (Section 1.6).
2. The Kalman filter is regarded as formidable by most engineers because it is traditionally expostulated in full-blown matrix form. This book introduces it in a very simple scalar context, where the basic strategy is transparent, as is its extension to the general case (Chapter 6).
3. The book is exceptional in that it distinguishes between the science of extracting statistical information from raw data (Chapter 3)—for example, a time series about which nothing is known a priori—and that of analyzing specific statistical models (Chapter 4). The former motivates the concepts of statistical spectral analysis (such as the Wiener–Khintchine theory), and the latter applies and interprets them in specific physical contexts.
4. The book's premise, throughout, is that new techniques are best introduced by specific, low-dimensional examples, rather than attempting to strive for generality at the outset; the mathematics is easier to comprehend and more enjoyable. Specific instances are the derivations of the Yule–Walker equations (Section 4.4), the normal equations (Section 5.4), and the Wiener filter (Section 5.9).

In short, this book is not comprehensive and not rigorous, but it is unique in its simplified approach to the subject. It "greases the skids" for students embarking on advanced reading, while it provides an adequate one-semester survey of random processes for the nonspecialist.

Supplementary material—selected answers, examples, exercises, insights, and errata—will be made available as they are generated, at the author's web site: http://ee.eng.usf.edu/people/snider2.html.

MATLAB® is a registered trademark of The MathWorks, Inc. For product information, please contact:

The MathWorks, Inc.
3 Apple Hill Drive
Natick, MA 01760-2098 USA
Tel: 508-647-7000
Fax: 508-647-7001
E-mail: info@mathworks.com
Web: www.mathworks.com

Author

Dr. Arthur David Snider has more than 50 years of experience in modeling physical systems in the areas of heat transfer, electromagnetics, microwave circuits, and orbital mechanics, as well as the mathematical areas of numerical analysis, signal processing, differential equations, and optimization.

Dr. Snider holds degrees in mathematics (BS from Massachusetts Institute of Technology and PhD from New York University) and physics (MA from Boston University), and he is a registered professional engineer. He served for 45 years on the faculties of mathematics, physics, and electrical engineering at the University of South Florida. He worked five years as a systems analyst at MIT's Draper Instrumentation Lab and has consulted in many industries in Florida. He has published seven textbooks in applied mathematics.

1 Probability Basics
A Retrospective

This textbook is intended for readers who have already studied basic statistics and probability. The purpose of Chapter 1 is to jog your memory by reviewing the material from a different and, hopefully, refreshing perspective.

1.1 WHAT IS "PROBABILITY"?

Since this isn't your first course in probability, you know how profound this question is. Let's consider a few ways in which we use the word.

There are some situations in which there is no question about the numerical values of the probabilities. If a bowl contains seven red balls and three green balls and one is picked at random, the probability that it is green is 0.3. It's hard to argue with that—although I might be backed into a corner if you asked me to define "at random." I might parry by saying each ball is "equally likely" or "equally probable." And then you've got me over a barrel, because my definition is circular.

But what if I select a ball "at random," note that it is red, and wrap my fist around it without showing it to you? To you, the probability of green is 0.3. But I *know* the color of the ball, and to me the probability of green is 0. Probability can be subjective; it is a measure of our state of knowledge. In fact the probability of green, from your point of view, will jump from 0.3 to 0 if I open my hand.

Some would argue that the probability is either 0 or 1 after the ball was drawn, whether or not I looked, because there is no randomness after the selection, only insufficient knowledge. In fact, there are some situations in quantum mechanics* where the possible physical outcomes of experiments are predicted to be different, according to whether or not the experimenter looks at the dials.

So let's try to bypass this subjective "insider information" aspect of probability, and consider the *long-term-average* interpretation. One is inclined to say that if the pick-a-ball experiment is performed many times (replacing the ball after recording the result), 30% of the times the ball will be green.

Is that true? Of course not! No matter how many repetitions are run, it is always *possible* that 29%, 32%, or 99% of the balls selected will be green. Indeed, the science of probability enables us to quantify how unlikely such results are.

But what, after all, does "unlikely" mean? We're still going around in circles.

Equally devastating for this interpretation are such everyday statements such as "there was a 30% chance of rain yesterday." How do we assess such a claim?

---

* Einstein, A., Podolsky, B., and Rosen, N. 1935. Can quantum-mechanical description of physical reality be considered complete? *Phys. Rev.* 41: 777. See also Bohm, D. 1957. *Quantum Theory.* New York: Dover; and Herbert, N. 1987. *Quantum Reality.* New York: Anchor.

We can't run a series of experiments where we replicate "yesterday" 500 times and calculate the number of times it rains. Either it rained or it didn't.

So probability is an elusive concept, although it has been explicated by philosophers and axiomatized by mathematicians. Nonetheless we feel comfortable discussing it, calculating it, and assessing strategies using it. Personally, I feel more comfortable when I can visualize a long-term-average interpretation for an answer that I have computed, even though I acknowledge the fallacies involved.

Let me give you an example. This problem has always been popular with textbook authors, but it achieved notoriety in the movie "21" in 2007.

A game show host tells you that a prize lies behind one of three doors, and you will win it if you select the correct door. The probability that you win is, of course, 1/3.

But then an intriguing strategy is added. After you have made a choice but before the chosen door is opened, the host (who knows where the prize is) opens one of the unchosen doors *not* containing the prize. Has this additional information—that the revealed door does not contain the prize—changed the probability that you have won? Has it increased to 1/2 (because now there is one choice out of 2)?

Should you change your choice of doors?

Bayes' theorem in Section 1.3 handles these questions, but let's try to find a long-time-average interpretation. So suppose every day for the next 20 years you play this game. You expect to pick the correct door 1/3 of those times. Regardless of whether you chose correctly or not, there is always an "empty" door for the host to open, so his action is not going to change whether or not you were correct; your probability of choosing the right door remains 1/3.

So 2/3 of the time you have picked a wrong door. What happens when you pick a wrong door? In such a case the host's hands are tied. He can't "choose" which door to open—he *has* to open the one that does not contain the prize. In other words, he's giving away the secret when he opens the second door!

Therefore, in 2/3 of the trials the prize is (for certain) behind the door that the host does not open, and in the other 1/3 of the trials, it is behind the door that you originally chose. If you decide to change your choice to the unopened door, you will win every time your initial choice was wrong; your probability of winning increases—not to 1/2, but to 2/3. Dang!

So even though the long-time-average interpretation of probability is logically untenable, we feel comfortable with it, and in this book I will not pursue the *definition* of probability further. In most applications, one *postulates* probabilities for a certain collection of outcomes (perhaps "equally likely"), and our mathematical theory describes how to compute other probabilities from them.

ONLINE SOURCES

The game show problem was posed in Marilyn vos Savant's column in *Parade* magazine in 1990. *I guarantee* you will be entertained if you read her account at vos Savant, Marilyn. Game show problem. *Parade Magazine*. Accessed June 24, 2016. http://marilynvossavant.com/game-show-problem/.

The relative-frequency interpretation of probability is explored through simulations at Boucher, Chris. The frequentist intuition behind assigning probabilities. Wolfram

Demonstrations Project. Accessed June 24, 2016. http://demonstrations.wolfram.com/ TheFrequentistIntuitionBehindAssigningProbabilities/.

A good source for the jargon employed by statisticians is

National Institute of Standards and Technology. Quantile-quantile plot. *Engineering Statistics Handbook*. Accessed June 24, 2016. http://www.itl.nist.gov/div898/handbook/eda/ section3/qqplot.htm.

1.2 THE ADDITIVE LAW

Figure 1.1 displays a standard rack of billiard balls and the sky. Suppose one ball is selected at random from the rack.

We could pose such questions as the following:

1. What is the probability that the 8-ball is chosen?
2. What is the probability that a striped ball is chosen?
3. What is the probability that a chosen ball contains one of the colors of the sky?

We have here a collection of 15 "basic" possibilities—the 1-ball is chosen, the 2-ball is chosen, etc.—and in the absence of any other information, we presume they are equally likely; the probability that the 8-ball is chosen, for example, is 1/15. The probability that any compound statement such as (2) or (3) is true is the sum of the probabilities of those basic events that make the statement true:

1. Probability of the 8-ball = 1/15 because only one ball lies in the statement's "truth set."
2. The striped balls, #9 through #15, comprise the truth set and the probability is 7/15.
3. Balls #2 and #9 through #15 contain blue or white (ball #10 contains both), so they comprise the truth set and the probability is 8/15.

FIGURE 1.1   Billiard balls and sky.

In general, we see that one way to visualize the laws of probability is to presume that there is a "universe" U of elemental distinct events whose probabilities are postulated, and we seek to assess the probabilities of compound statements that define specific *subsets* of U by summing the elemental probabilities of the events in the subset. See Figures 1.2 and 1.3.

U
```
x  x  x  x  x  x  x  x  x  x  x  x  x  x

x  x  x  x  x  x  x  x  x  x  x  x  x  x

x  x  x  x  x  x  x  x  x  x  x  x  x  x

x  x  x  x  x  x  x  x  x  x  x  x  x  x

x  x  x  x  x  x  x  x  x  x  x  x  x  x

x  x  x  x  x  x  x  x  x  x  x  x  x  x

x  x  x  x  x  x  x  x  x  x  x  x  x  x

x  x  x  x  x  x  x  x  x  x  x  x  x  x
```

FIGURE 1.2   The universe of elemental events.

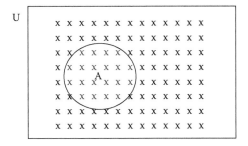

FIGURE 1.3   The truth set of statement A.

Then from Figure 1.4, it is easy to ascertain the validity of the following statement.

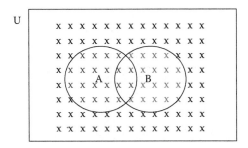

FIGURE 1.4   Additive law of probability.

The Additive Law of Probability: The probability of A or B equals the probability of A plus the probability of B minus the probability of both A and B.

Employing the "union" and "intersection" symbols from set theory, we express the additive law as

$$p(A \cup B) = p(A) + p(B) - p(A \cap B). \tag{1.1}$$

Thus for statement (3), we reason as follows:

The probability that the chosen ball contains either the color blue or the color white

equals

the probability that it contains the color blue (2/15: 2-ball and 10-ball)

plus

the probability that it contains the color white (7/15: balls #9 through #15)

minus

the probability that it contains both blue and white (1/15: the 10-ball).

$$\frac{2}{15} + \frac{7}{15} - \frac{1}{15} = \frac{8}{15}.$$

The negative term accounts for the fact that the 10-ball was counted twice.

1.3 CONDITIONAL PROBABILITY AND INDEPENDENCE

Consider the following questions:

1. What is the probability that a chosen billiard ball's number is a prime (2,3,5,7,11,13)?
2. What is the probability that a chosen ball's number is a prime (2,3,5,7,11,13), *given that the ball is striped*?

Of course the answer to (1) is simply 6/15; there are 15 equally likely choices, and 6 of them comprise the truth set of the statement "the ball's number is a prime." But since we are told in question (2) that the chosen ball is striped, our universe of elemental events is altered; it only contains balls #9 through #15. Thus there are now seven equally likely events (consult Figure 1.1), so their probabilities must be "renormalized" to 1/7 each. The prime-numbered balls in this new universe are #11 and #13, so the answer to (2) is 2/7.

Questions like (2) address *conditional probabilities*; a condition is stated that alters the universe of possible events. In this context, questions like (1), that contain no conditions, are called *a priori probabilities*. Let us generalize the reasoning in the preceding paragraph to deduce the formulas that relate conditional and a priori probabilities.

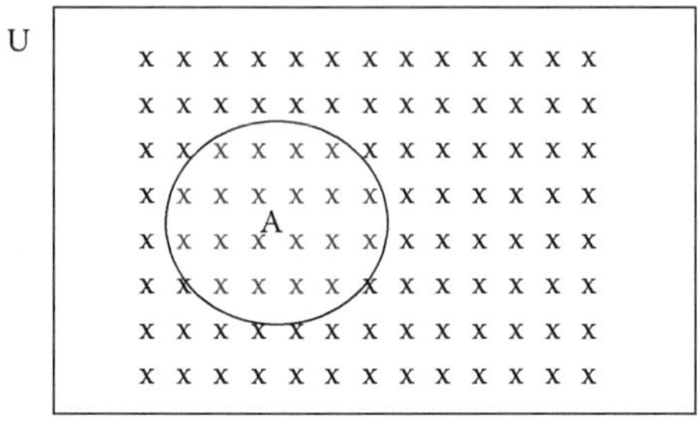

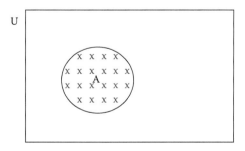

FIGURE 1.5   Universe.

In Figure 1.5 focus only on the elemental events that make up the truth set for the statement A; the (a priori) probability of A is the sum of the probabilities of these events:

$$p(A) = \sum_{x \in A} p(x),$$ (1.2)

where "$x \in A$" means "x is in the truth set of A." But if we are given that A is true, the universe of elemental events collapses onto the truth set of A, as in Figure 1.6. The conditional probability of A, *given A*, is obviously 1, and we express this by writing

$$p(A|A) = \sum_{x \in A} p(x|A) = 1,$$ (1.3)

using the notation "...|A" to denote "..., given A."

FIGURE 1.6   Conditional universe, given A.

What are the elemental *conditional* probabilities $p(x|A)$? They should retain the same relative proportions; for example, if x_1 and x_2 are each in the truth set of A and they are equally likely *a priori*, then we assume they will still be equally likely when A is known to be true. So all of these probabilities $p(x)$ should be rescaled by the same factor when A is given. And when you reexpress (1.2) as

$$\sum_{x \in A} \frac{p(x)}{p(A)} = 1$$ (1.4)

and compare with (1.3), it's obvious this constant factor is $1/p(A)$:

$$p(x|A) = \frac{p(x)}{p(A)} \quad \text{if } x \in A \left(\text{and of course } p(x|A) = 0 \text{ otherwise}\right). \quad (1.5)$$

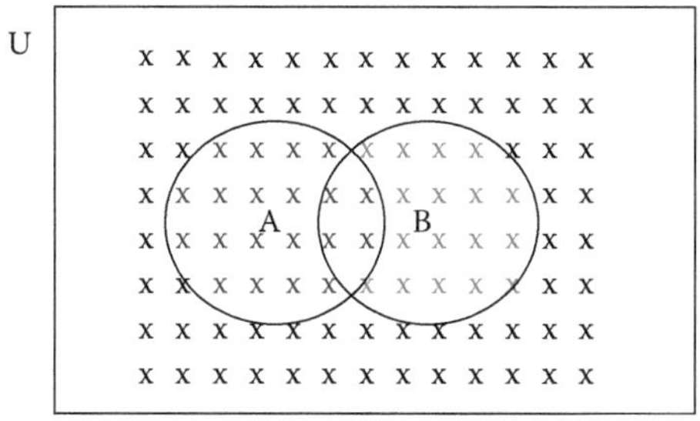

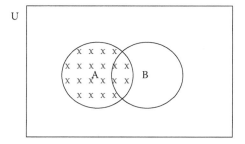

FIGURE 1.7 *A priori* universe.

FIGURE 1.8 Conditional universe.

Referring to Figures 1.7 and 1.8, we can derive the equation for the conditional probability of any statement B, given A, by restricting to the "conditional universe" and totaling the conditional probabilities of the elemental events in the truth set of B:

$$p(B|A) = \sum_{x \in A \cap B} p(x|A) = \sum_{x \in A \cap B} \frac{p(x)}{p(A)} = \frac{1}{p(A)} \sum_{x \in A \cap B} p(x) = \frac{p(A \cap B)}{p(A)}.$$

As a result $p(A \cap B) = p(A)p(B|A)$, and since $A \cap B = B \cap A$, we have the following:

Laws of Conditional Probability

$$p(A \cap B) = p(A)p(B|A) = p(B)p(A|B). \quad (1.6)$$

In particular,

$$p(B|A) = \frac{p(B)p(A|B)}{p(A)}. \quad (1.7)$$

(1.7) is known as **Bayes' theorem.**

To check our earlier calculation, we manipulate (1.6) to write

$$p(\text{prime number} \mid \text{striped ball}) = \frac{p(\text{prime number and striped ball})}{p(\text{striped ball})}$$

$$= \frac{p(\text{balls 11 or 13})}{p(\text{balls 9 through 15})}$$

$$= \frac{2/15}{7/15} = \frac{2}{7}.$$

Now if the probability of B is unaffected by the truth of A, then colloquially we say that B is "independent" of A. *Probabilistic independence* is thus characterized by the identity

$$p(B \mid A) = p(B) \quad \text{if B is independent of A.}$$

Note that this is equivalent to saying that A is independent of B (see 1.7) and that (1.6) becomes

$$p(A \cap B) = p(A) p(B)$$

when A and B are independent. To summarize:

Independence: The statements A and B are independent when any of the following equivalent conditions are true:

$$p(B \mid A) = p(B). \tag{1.8}$$

$$p(A \mid B) = p(A). \tag{1.9}$$

$$p(A \cap B) = p(A) p(B). \tag{1.10}$$

Although formulas (1.8) or (1.9) best convey the notion of independence, criterion (1.10) is often the most convenient for applications.

SUMMARY: IMPORTANT LAWS OF PROBABILITY

Additive law	$p(A \cup B) = p(A) + p(B) - p(A \cap B).$
Conditional probability	$p(A \cap B) = p(A)p(B \mid A) = p(B)p(A \mid B).$
Bayes' theorem	$p(A \mid B) = \dfrac{p(B \mid A) p(A)}{p(B)}.$
Independence	$p(A \cap B) = p(A)p(B).$

1.4   PERMUTATIONS AND COMBINATIONS

The formulas for the number of permutations and combinations of a set of objects play an important role in the applications of statistics, so we devote this section to a review of the concepts.

How many ways are there of ordering the set of numbers {1, 2, 3, 4}? There are four candidates for the first slot:

$$\begin{bmatrix} 1 & \times & \times & \times \end{bmatrix} \begin{bmatrix} 2 & \times & \times & \times \end{bmatrix} \begin{bmatrix} 3 & \times & \times & \times \end{bmatrix} \begin{bmatrix} 4 & \times & \times & \times \end{bmatrix}$$

For each of these assignments, there are three possibilities for the second slot:

$$\begin{bmatrix} 1 & 2 & \times & \times \end{bmatrix} \begin{bmatrix} 2 & 1 & \times & \times \end{bmatrix} \begin{bmatrix} 3 & 1 & \times & \times \end{bmatrix} \begin{bmatrix} 4 & 1 & \times & \times \end{bmatrix}$$
$$\begin{bmatrix} 1 & 3 & \times & \times \end{bmatrix} \begin{bmatrix} 2 & 3 & \times & \times \end{bmatrix} \begin{bmatrix} 3 & 2 & \times & \times \end{bmatrix} \begin{bmatrix} 4 & 2 & \times & \times \end{bmatrix}$$
$$\begin{bmatrix} 1 & 4 & \times & \times \end{bmatrix} \begin{bmatrix} 2 & 4 & \times & \times \end{bmatrix} \begin{bmatrix} 3 & 4 & \times & \times \end{bmatrix} \begin{bmatrix} 4 & 3 & \times & \times \end{bmatrix}$$

For each of these, there are two candidates for the third slot (and the final slot is filled by default):

$$\begin{bmatrix} 1 & 2 & 3 & 4 \end{bmatrix} \begin{bmatrix} 2 & 1 & 3 & 4 \end{bmatrix} \begin{bmatrix} 3 & 1 & 2 & 4 \end{bmatrix} \begin{bmatrix} 4 & 1 & 2 & 3 \end{bmatrix}$$
$$\begin{bmatrix} 1 & 2 & 4 & 3 \end{bmatrix} \begin{bmatrix} 2 & 1 & 4 & 3 \end{bmatrix} \begin{bmatrix} 3 & 1 & 4 & 2 \end{bmatrix} \begin{bmatrix} 4 & 1 & 3 & 2 \end{bmatrix}$$
$$\begin{bmatrix} 1 & 3 & 2 & 4 \end{bmatrix} \begin{bmatrix} 2 & 3 & 1 & 4 \end{bmatrix} \begin{bmatrix} 3 & 2 & 1 & 4 \end{bmatrix} \begin{bmatrix} 4 & 2 & 1 & 3 \end{bmatrix}$$
$$\begin{bmatrix} 1 & 3 & 4 & 2 \end{bmatrix} \begin{bmatrix} 2 & 3 & 4 & 1 \end{bmatrix} \begin{bmatrix} 3 & 2 & 4 & 1 \end{bmatrix} \begin{bmatrix} 4 & 2 & 3 & 1 \end{bmatrix}$$
$$\begin{bmatrix} 1 & 4 & 2 & 3 \end{bmatrix} \begin{bmatrix} 2 & 4 & 1 & 3 \end{bmatrix} \begin{bmatrix} 3 & 4 & 1 & 2 \end{bmatrix} \begin{bmatrix} 4 & 3 & 1 & 2 \end{bmatrix}$$
$$\begin{bmatrix} 1 & 4 & 3 & 2 \end{bmatrix} \begin{bmatrix} 2 & 4 & 3 & 1 \end{bmatrix} \begin{bmatrix} 3 & 4 & 2 & 1 \end{bmatrix} \begin{bmatrix} 4 & 3 & 2 & 1 \end{bmatrix}$$

$$(1.11)$$

In all, then, there are $4 \times 3 \times 2 = 4!$ orderings or *permutations* of four objects. In general:

Permutations: The number of permutations of n objects equals $n!$.

Example 1.1: There are about one trillion ways of lining up all 15 billiard balls: $15! \approx 1.3077 \times 10^{12}$.

Now we ask, how many ways are there for selecting *pairs* from the set {1 2 3 4}, or "how many combinations are there among four items, taking two at a time?" The answer is six, namely, (1,2), (1,3), (1,4), (2,3), (2,4), and (3,4). How can we verify and generalize this?

We can visualize the pairs by studying the permutation display (1.11). Imagine a partition, separating the first two numbers in each foursome from the others; thus

$\{3 \quad 1 \; : \; 2 \quad 4\}$ designates the selection of the pair consisting of 3 and 1 (with 2 and 4 left over). But this particular pairing appears more than once in the display,

$$
\begin{array}{cccc}
\begin{bmatrix} 1 & 2 & : & 3 & 4 \end{bmatrix} & \begin{bmatrix} 2 & 1 & : & 3 & 4 \end{bmatrix} & *\begin{bmatrix} 3 & 1 & : & 2 & 4 \end{bmatrix}* & \begin{bmatrix} 4 & 1 & : & 2 & 3 \end{bmatrix} \\
\begin{bmatrix} 1 & 2 & : & 4 & 3 \end{bmatrix} & \begin{bmatrix} 2 & 1 & : & 4 & 3 \end{bmatrix} & *\begin{bmatrix} 3 & 1 & : & 4 & 2 \end{bmatrix}* & \begin{bmatrix} 4 & 1 & : & 3 & 2 \end{bmatrix} \\
\begin{bmatrix} 1 & 3 & : & 2 & 4 \end{bmatrix} & \begin{bmatrix} 2 & 3 & : & 1 & 4 \end{bmatrix} & \begin{bmatrix} 3 & 2 & : & 1 & 4 \end{bmatrix} & \begin{bmatrix} 4 & 2 & : & 1 & 3 \end{bmatrix} \\
\begin{bmatrix} 1 & 3 & : & 4 & 2 \end{bmatrix} & \begin{bmatrix} 2 & 3 & : & 4 & 1 \end{bmatrix} & \begin{bmatrix} 3 & 2 & : & 4 & 1 \end{bmatrix} & \begin{bmatrix} 4 & 2 & : & 3 & 1 \end{bmatrix} \\
\begin{bmatrix} 1 & 4 & : & 2 & 3 \end{bmatrix} & \begin{bmatrix} 2 & 4 & : & 1 & 3 \end{bmatrix} & \begin{bmatrix} 3 & 4 & : & 1 & 2 \end{bmatrix} & \begin{bmatrix} 4 & 3 & : & 1 & 2 \end{bmatrix} \\
\begin{bmatrix} 1 & 4 & : & 3 & 2 \end{bmatrix} & \begin{bmatrix} 2 & 4 & : & 3 & 1 \end{bmatrix} & \begin{bmatrix} 3 & 4 & : & 2 & 1 \end{bmatrix} & \begin{bmatrix} 4 & 3 & : & 2 & 1 \end{bmatrix}
\end{array}
$$

$$(1.12)$$

because in counting combinations, we do not distinguish the ordering *within* the selected pair (or within the leftovers). In fact since the pair (1, 3) can be ordered 2! ways and the leftovers (2, 4) can also be ordered 2! ways, the selection of this pair is listed $(2!)(2!) = 4$ times in the permutation display (1.12). Indeed, *every* pair appears $(2!)(2!)$ times. So the total number of different pairs equals $4!/(2!2!) = 6$.

By the same reasoning, every *triplet* such as $\{2\ 3\ 4 : 1\}$ appears $(3!)(1!)$ times in the display, so there are $4!/(3!1!) = 4$ different triplets: specifically, $\{1\ 2\ 3\}$, $\{2\ 3\ 4\}$, $\{1\ 2\ 4\}$, and $\{2\ 3\ 4\}$. (And the same number of singletons, since every partition into a triplet and a leftover singleton corresponds to a partition into a singleton and a leftover triplet.) The generalization is as follows:

Combinations: The number of combinations of n objects, taking n_1 at a time, equals

$$
\frac{n!}{n_1!\,(n - n_1)!}.
$$

Example 1.2: Among a set of 15 unpainted billiard balls, there are 6,435 ways of selecting the 7 balls to be painted with stripes: $15!/(7!8!) = 6{,}435$.

We can generalize this. Suppose we partitioned 15 billiard balls into groups of 3, 5, and 7 balls. A typical partition would look like

$$
\{2\ 4\ 5 : 10\ 11\ 13\ 14\ 15 : 1\ 3\ 6\ 7\ 8\ 9\ 12\},
$$

and it would appear in the permutation list $(3!)(5!)(7!)$ times. Therefore the number of possible such partitions equals $15!/(3!5!7!) = 360{,}360$.

Partitions: The number of ways of partitioning n objects into sets of size n_1, $n_2, \ldots, n_k$

$$
\left(n = n_1 + n_2 + \cdots + n_k\right) \quad equals \quad \frac{n!}{n_1!\,n_2! \cdots n_k!}.
$$

1.5   CONTINUOUS RANDOM VARIABLES

Although we didn't say so explicitly, the number of elemental events in the "universe" for any particular probabilistic experiment has been assumed thus far to be finite. However, for an experiment such as selecting a real number x from an infinite number of possibilities—the *interval* between −1 and 1, say—we have to be a little more imaginative with the mathematics. If every real number in the interval (−1, 1) is equally likely, then we reach the inescapable conclusion that the probability of each number is zero; any *non*zero probability, accumulated over the infinite population of (−1, 1), would give a net probability of infinity.

It is clear, however, that the probability that $X > 0$ must be 1/2. And the probability that $X < −1/2$ must be 1/4. So we postulate that the probability is distributed over the interval (−1, 1) with a continuous *density* $f_X(x)$ as depicted in Figure 1.9, and the probability that X lies between a and b equals the area under the curve $y = f_X(x)$ between a and b.

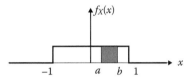

FIGURE 1.9   Probability density function.

Since the total area under $f_X(x)$ must then be one, the height of the curve is 1/2.

If we propose that numbers between 0 and 1 are twice as likely to occur as those between −1 and 0, then the situation looks like that in Figure 1.10.

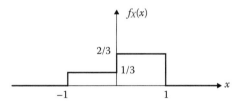

FIGURE 1.10   Skewed pdf.

A **probability density function** ("pdf"), then, is a function $f_X(x)$ defined for all real x that is never negative and has total integral equal to 1 (Figure 1.11).

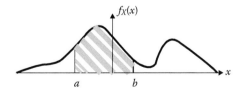

FIGURE 1.11 $f_X(x) \geq 0, \int_{-\infty}^{\infty} f_X(x)dx = 1.$

Thus for a continuous random variable X, the probability of a statement A equals the integral of the pdf, $f_X(x)$, over the truth set of A.

$$p(A) = \int_{x \in A} f_X(x)\, dx.$$

This makes the probability of a single point equal to zero. It also raises some interesting issues in set theory, which we defer to Section 1.6.

We need to introduce a little flexibility into the probability density function description, however, to handle some situations. Consider this experiment: We flip a coin, and if it comes up heads, we set $X = 0$ and the experiment is done, but if it comes up tails, X is selected at random to be a number between 1 and 4, each equally likely.

This is a situation where the probability of a single point, namely $X = 0$, is *not* zero; it is 1/2. The values of X between 1 and 4 have a net probability of 1/2, equally distributed over the interval [1, 4] by a probability density function of height 1/6 ($= 1/2 \div 3$), and we have no trouble graphing this part. But our efforts to pack an area of 1/2 over the single point $x = 0$ leads to futile sketches like those in Figure 1.12.

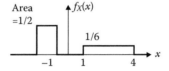

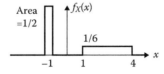

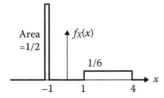

FIGURE 1.12   "Approximate" probability density functions.

What we would like is to have a function that is zero everywhere except for one point, where it is infinite, but with its integral finite (and nonzero). Oliver Heaviside first proposed expanding the calculus to include such "generalized functions." (Paul Dirac used them extensively in his formulation of quantum mechanics,

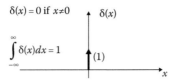

FIGURE 1.13   Delta function.

and Laurent Schwartz devised the mathematical interpretation that enables their rigorous implementation.*,†,‡)

The **delta function** $\delta(x)$ is depicted in Figure 1.13; it is an "impulse" supported at $x = 0$, spanning unit area.

Refer to Figure 1.13 to make sure you understand the following properties:

$$f(x)\delta(x) \equiv f(0)\delta(x).$$

$$f(x)\delta(x - x_0) \equiv f(x_0)\delta(x - x_0).$$

$$\int_{-\infty}^{\infty} f(x)\delta(x - x_0)\,dx = \int_{-\infty}^{\infty} f(x_0)\delta(x - x_0)\,dx = f(x_0).$$

Using the delta function, we can accommodate discrete values of X that have positive probabilities into the probability density function description. The pdf that we were trying to depict in Figure 1.12 is described by

$$f_X(x) = \frac{1}{2}\delta(x) + \begin{cases} 1/6 & \text{if } 1 < x < 4, \\ 0 & \text{otherwise.} \end{cases}$$

(see Figure 1.14).

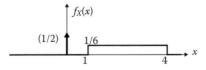

FIGURE 1.14   Mixed discrete and continuous pdf.

---

* Berg, E.J. 1936. *Heaviside's Operational Calculus*, 2nd ed. New York: McGraw Hill.
† Medvedev, F.A. 1988. The delta-function of G L Giorgi and P A M Dirac (Russian). In *Studies in the History of Physics and Mechanics*. Moscow: Nauka.
‡ Schwartz, L. 1966. *Theorie des distributions*. Paris: Hermann.

FIGURE 1.15   Mixed discrete and continuous pdf.

A situation where the pdf is continuous except for *two* discrete points having probabilities 1/4 each is depicted in Figure 1.15.

The use of delta functions can be avoided by employing the notion of the **cumulative distribution function** $F_X(x)$, which equals the probability that the random variable X is less than the number x. In other words,

$$F_X(x) = \int_{-\infty}^{x} f_X(\xi)d\xi.$$

(If the pdf contains a delta function supported at x_0 we integrate up to $(x_0 - \varepsilon)$ and take the limit as $\varepsilon{\downarrow}0$.)

Note the details of how $F_X(x)$ is related to $f_X(x)$ in Figure 1.16—in particular, note how the improprieties of the delta functions are finessed by the cumulative distribution function formulation.

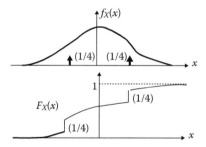

FIGURE 1.16   Cumulative distribution function.

Even though the cumulative distribution function fits more comfortably into rigorous calculus than the pdf, with its delta functions, the use of the former is cumbersome when we are formulating practical problems. Therefore, we shall seldom use the cumulative distribution function any further in this book.

Moreover, we will use the customary "infinitesimals" vernacular when describing probabilities with a continuous pdf. Therefore, we shall say

$f_X(x)\,dx$ equals the probability that X lies in the infinitesimal interval $x < X < x + dx$.

This is a shorthand for a long, verbose statement. The probability that X lies between a and b is the integral $\int_{a}^{b} f_X(\xi)d\xi$, which is the limit of the mathematician's "Riemann sums" $\sum_{i=1}^{n} f(x_i)\Delta x_i$ as each $\Delta x_i \to 0$. By writing "dx" instead of

"Δx," we acknowledge that we are skimming over the details of a rigorous argument involving the analysis of limits. If $f_X(x)$ has discontinuities—particularly delta functions—we should use this shorthand cautiously.

The pdf formulation of conditional probability for continuous random variables is easily derived using this informal language. Recall from the law of conditional probability in Section 1.3:

Probability that X lies between x and $x + dx$, given that A is true
equals

Probability that A is true and X lies between x and $x + dx$
divided by

Probability that A is true (1.13)

We denote the **conditional probability density function for X given A** by $f_{X|A}(x)$, so that the first quantity is $f_{X|A}(x)\, dx$. If x is in the truth set of A, then the second quantity is $f_X(x)\, dx$; otherwise it is zero. Thus (1.13) states

$$f_{X|A}(x)\,dx = \begin{cases} \dfrac{f_X(x)\,dx}{p(A)} & \text{if } x \text{ is in the truth set of } A, \\ 0 & \text{otherwise.} \end{cases} \tag{1.14}$$

For example, suppose A is the statement "$x > 5$." Then (dropping the common factor dx),

$$f_{X|A}(x) = \begin{cases} \dfrac{f_X(x)}{\displaystyle\int_5^\infty f_X(\xi)\,d\xi} & \text{if } x > 5, \\ 0 & \text{otherwise.} \end{cases}$$

SUMMARY: IMPORTANT FACTS ABOUT CONTINUOUS RANDOM VARIABLES

$f_X(x)$ is the probability density function ("pdf").

$f_X(x)\, dx$ equals the probability that X lies in the infinitesimal interval $x < X < x + dx$.

The delta-function property: $\displaystyle\int_{-\infty}^{\infty} f(x)\delta(x-x_0)\,dx = \int_{-\infty}^{\infty} f(x_0)\delta(x-x_0)\,dx = f(x_0)$

Cumulative distribution function: $F_X(x) = \displaystyle\int_{-\infty}^{x} f_X(\xi)\,d\xi$

Conditional pdf:

$$f_{X|A}(x) = \begin{cases} \dfrac{f_X(x)}{p(A)} & \text{if } x \text{ is in the truth set of } A, \\ 0 & \text{otherwise.} \end{cases}$$

1.6   COUNTABILITY AND MEASURE THEORY

For a continuous probability density function like that in Figure 1.9 of the preceding section, we have an apparently contradictory situation:

> The total probability of getting a value of X between -1 and 1 is unity.
> The probability of getting any specific value between -1 and 1 is zero.
> Probability is additive.

A rigorous foundation for probability theory must address this issue. Here we shall simply highlight some of the ideas, to facilitate your reading when you study more rigorous texts.

We begin by scrutinizing the question as to which of two sets, A or B, has more elements. It may seem trivial—we compare N_A, the number of elements in A, to the number N_B in B; that is, the set having the greater "cardinality" has more elements.

But this won't work if both sets are infinite. So we revert to a more primitive, "stone-age" test: match up the two sets element by element, and the set with elements left over is the greater. More precisely, if the elements of A can be matched one-to-one with a proper *sub*set of the elements of B, then the cardinality of B is greater.

It may come as a surprise to realize that this won't work either. Did you know that infinite sets can be put into one-to-one correspondences with subsets of themselves? For example, consider the set of positive integers. It would seem that its cardinality would be greater than that of the subset consisting of positive *even* numbers. Yet by doubling each integer in the former set we obtain a perfect one-to-one matching of the two sets, with no leftovers. They have the same cardinality.

So do all infinite sets have the same cardinality? No. It is natural to call a set that can be matched up one to one with the positive integers "countable," but the continuum of all real numbers in, say, the interval between 0 and 1 is *not* countable. To show this, identify each real number with its infinite decimal expansion:

$$1/2 = 0.50000000000000....$$
$$1/3 = 0.333333333333333....$$
$$\sqrt{2}/2 = 0.707106781186548....$$
$$\pi/4 = 0.785398163397448....$$

(In the interest of mathematical precision, two comments should be made as follows:

1. According to classical mathematical analysis, every infinite decimal can be interpreted as the limit of a convergent sequence and thus corresponds to a real number.
2. Any number, like 1/2, with a *terminating* decimal representation, has another, nonterminating representation exemplified by $0.4999999999999... = 0.5$, so we adopt the convention that only the terminating representation be used.)

To show that the continuum is not countable, suppose otherwise. Then the correspondence between the integers and the continuum could be visualized as in Figure 1.17.

#1	0.5000000000. . .
#2	0.33333333333. . .
#3	0.70710678118. . .
#4	0.78539816339. . .
. . .	. . .
#n	.*xxxxxxxxxxx*. . .
. . .	. . .

FIGURE 1.17   Proposed enumeration of (0,1).

Now form a decimal number by taking the first digit from entry #1 (i.e., "5"), and change it to something else (e.g., "6"); take the second digit from entry #2 ("3") and change it to something else ("4"); take the third digit from entry #3 and change it, and so on (taking care to avoid an infinite run of 9's).

This new number is not in the list, because it differs from #1 in the first digit, from #2 in the second digit, from #3 in the third digit, and so on. In other words, a *countable* list cannot possibly include every real number in the interval, so the cardinality of the latter exceeds the cardinality of the integers.

This raises some interesting set-theoretic questions*:

1. Is there a hierarchy of ever-higher cardinalities? (Answer: yes.)
2. Is the cardinality of the *rational* numbers higher than that of the integers? (Answer: no.)
3. Is the cardinality of the real numbers in the two-dimensional plane higher than that of those on the one-dimensional line? (Answer: no.)
4. Is there a set whose cardinality lies between that of the integers and that of the continuum? (Answer: This is called the "continuum hypothesis," and its answer is controversial.)

But this exploration suggests how to reconcile the observations highlighted at the beginning of this section: The total probability for all the numbers in the continuous interval (0,1) is one. The probability of any individual number in the continuum is zero. The continuum is the union of its individual numbers. Does the additive law claim we can get one by adding up zeros?

---

* Tiles, M. 2004. *The Philosophy of Set Theory: An Historical Introduction to Cantor's Paradise.* New York: Dover.

Observing the distinctions between different "orders of infinity," theorists devised a fix up to the laws of probability that, at bottom, restricts the additive law to *countable* unions. The interval (0, 1) is the union of the intervals [1/2, 1) plus [1/4, 1/2) plus [1/8, 1/4) plus ... plus [$1/2^n$, $1/2^{n-1}$), ... And if we add up the probabilities of each of these (disjoint) subintervals, the sum of the resulting series converges to 1.

But since the continuum is *not* the union of a countable number of points, the restriction means there will be no probabilistic argument that is going to assert that 1 is a sum of zeros.

When you read a rigorous book on probability, you will see that probabilities are not assigned to every subset of the continuum, but rather to the so-called "measureable" sets; these include individual points, closed intervals [a,b], open intervals (c,d), and—most significantly—*countable* unions and intersections of these; the jargon "sigma-algebras" is employed. The probability density functions respect this structure; they are said to be "measurable functions," and their integrals are defined over measurable sets.

Do not be intimidated by this parade of mathematical rigor. It is just a sagacious way of avoiding the paradoxical traps we have seen.

1.7 MOMENTS

If X is a random variable and $g(x)$ is any function, the **Expected Value** of $g(X)$, denoted both by $E\{g(X)\}$ and $\overline{g(X)}$, is the weighted average of the values of $g(X)$, weighted by the probability density function:

$$\int_{-\infty}^{\infty} g(x) f_X(x)\, dx = \overline{g(X)} = E\{g(X)\}. \tag{1.15}$$

For example, the expected value of $\sin X$ is $\overline{\sin X} = E\{\sin X\} = \int_{-\infty}^{\infty} \sin x\, f_X(x)\, dx$. Note two facts:

1. The expected value is a linear function:

$$E\{\alpha g(X) + \beta h(X)\} = \alpha E\{g(X)\} + \beta E\{h(X)\}. \tag{1.16}$$

2. The expected value of any constant is the constant, itself (since $\int_{-\infty}^{\infty} f_X(x)\, dx = 1$).

The **moments** of X are the weighted averages of its powers.

$$n\text{th moment of } X = \int_{-\infty}^{\infty} x^n f_X(x)\, dx = \overline{X^n} = E\{X^n\}. \tag{1.17}$$

The first moment is the **mean** of X and is denoted μ_X:

$$\int_{-\infty}^{\infty} x f_X(x) dx = \mu_X = \overline{X} = E\{X\}. \qquad (1.18)$$

μ_X locates the "center of mass" of the pdf. Remember that μ_X is a constant, not a random variable. (So its expected value is—what?)

The second moment of X is its **mean square**:

$$\int_{-\infty}^{\infty} x^2 f_X(x) dx = \overline{X^2} = E\{X^2\}. \qquad (1.19)$$

Despite the similarity in notation, $\overline{X^2}$ is a very different object from the square of the mean $\overline{X}^2$.

If we take the expected value of the deviation of X from its mean, we would expect to get zero since the positive values of $(X - \mu_X)$ cancel the negative values, on the average:

$$E\{X - \mu_X\} = \int_{-\infty}^{\infty} (x - \mu_X) f_X(x) dx = E\{X\} - E\{\mu_X\} = \mu_X - \mu_X = 0.$$

So to get a true measure of the spread of X from its mean, we take the expected value of $(X - \mu_X)^2$ (which is always nonnegative), and then take the square root. This "root-mean-square" of $(X-\mu_X)$ is called the **standard deviation** σ_X:

$$\sigma_X = \sqrt{\int_{-\infty}^{\infty} (x - \mu_X)^2 f_X(x) dx} = \sqrt{E\{(X - \mu_X)^2\}}. \qquad (1.20)$$

You should confirm the values cited in Figure 1.18 for the standard deviation, and convince yourself that in each case $2\sigma_X$ gives an approximate measure of the width of the pdf curve (σ_X to the right plus σ_X to the left).

The square of the standard deviation is known as the **variance**, but it doesn't have a separate symbol (we simply use σ_X^2). The variance and the first and second moments are related.

Second-Moment Identity

$$\overline{X^2} = \overline{X}^2 + \sigma_X^2 \equiv \mu_X^2 + \sigma_X^2.$$

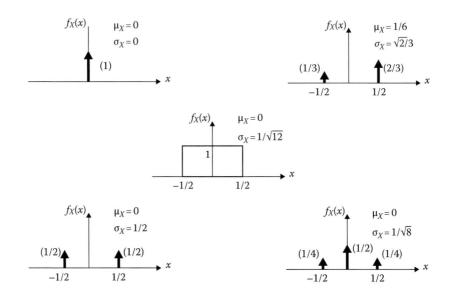

FIGURE 1.18 μ_X and σ_X.

This identity follows easily from the linearity property:

$$\sigma_X^2 = E\left\{\left(X-\mu_x\right)^2\right\} = E\left\{X^2 - 2X\mu_X + \mu_X^2\right\}$$
$$= E\left\{X^2\right\} - 2E\left\{X\right\}\mu_X + \mu_X^2$$
$$= \overline{X^2} - 2\mu_X\mu_X + \mu_X^2 = \overline{X^2} - \mu_X^2.$$

The inverse Fourier transform of the pdf $f_X(x)$ is called the **characteristic function** $\Phi_X(\omega)$ of X:

$$\Phi_X\left(\omega\right) = \int_{-\infty}^{\infty} e^{i\omega x} f_X\left(x\right)dx. \tag{1.21}$$

It is amusing to note that $\Phi_X(\omega)$ is, in fact, an expected value: $\Phi_X(\omega)=E\{e^{i\omega X}\}$. It is more productive, however, to look at the derivatives of $\Phi_X(\omega)$ at $\omega = 0$:

$$\Phi_X\left(\omega\right)\Big|_{\omega=0} = \int_{-\infty}^{\infty} e^{i\omega x} f_X\left(x\right)dx\Big|_{\omega=0} = 1; \tag{1.22}$$

$$\Phi_X'\left(\omega\right)\Big|_{\omega=0} = \int_{-\infty}^{\infty} ixe^{i\omega x} f_X\left(x\right)dx\Big|_{\omega=0} = iE\left\{X\right\},$$

$$\Phi_X''(\omega)\Big|_{\omega=0} = \int_{-\infty}^{\infty}(ix)^2 e^{i\omega x}f_X(x)dx\Bigg|_{\omega=0} = i^2 E\{X^2\},$$

$$\Phi_X'''(\omega)\Big|_{\omega=0} = \int_{-\infty}^{\infty}(ix)^3 e^{i\omega x}f_X(x)dx\Bigg|_{\omega=0} = i^3 E\{X^3\},$$

and so on. *The derivatives of $\Phi_X(\omega)$ at $\omega=0$ are the moments of X* (up to trivial constants). So a table of Fourier transforms* is a source of examples of characteristic functions.

In this book, we will seldom have occasion to use $\Phi_X(\omega)$ directly, but we shall see that it is occasionally a valuable *theoretical* tool for deducing certain properties of X. For instance, already we are in a position to establish the surprising result that in many circumstances *a probability density function is completely determined by its moments*, $E\{X^n\}$. Why? The moments determine the derivatives of $\Phi_X(\omega)$ at $\omega = 0$; thus they determine the Maclaurin series for $\Phi_X(\omega)$, and hence $\Phi_X(\omega)$ itself if it is analytic, and $f_X(x)$ is given by the Fourier transform of $\Phi_X(\omega)$.[†]

SUMMARY: IMPORTANT FACTS ABOUT EXPECTED VALUE AND MOMENTS

Expected value: $\displaystyle\int_{-\infty}^{\infty} g(x)f_X(x)\,dx = \overline{g(X)} = E\{g(X)\}.$

Generating function: $\displaystyle\Phi_X(\omega) = \int_{-\infty}^{\infty} e^{i\omega x}f_X(x)dx = E\{e^{i\omega x}\}.$

Moments: $\displaystyle E\{X^n\} = \int_{-\infty}^{\infty} x^n f_X(x)\,dx = i^{-n}\Phi_X^{(n)}(\omega)\Big|_{\omega=0}$

$$E\{X\} = \mu_X = \overline{X}, \quad E\{X^2\} = \overline{X^2}, \dots$$

Standard deviation: $\displaystyle\sigma_X = \sqrt{\int_{-\infty}^{\infty}(x-\mu_X)^2 f_X(x)\,dx} = \sqrt{E\{(X-\mu_X)^2\}}.$

Second-moment identity: $\overline{X^2} = \mu_X^2 + \sigma_X^2.$

1.8 DERIVED DISTRIBUTIONS

Suppose X is a random variable with a pdf $f_X(x)$ and we need the pdf of a related random variable such as X^2 or e^X or, generically, $g(X)$. In Figure 1.19 we depict the relation between the new variable $Y = g(X)$ and the original X, the pdf $f_X(x)$, and the pdf $f_Y(y)$ for Y.

---

* Of nonnegative functions, of course.

† A more painstaking analysis reveals that some pdfs have characteristic functions that are *not* analytic at $\Phi_X(\omega)$—the lognormal distribution defined in Problem 27 is an example.

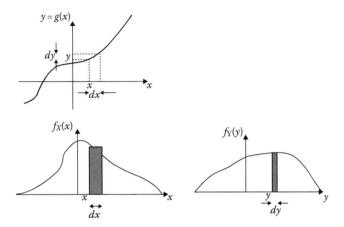

FIGURE 1.19 Change of (random) variable.

Since Y is related to X by $Y = g(X)$, the event "X lies between x and $x + dx$" is precisely the same as the event "Y lies between y and $y + dy$" when $y = g(x)$ and $y + dy = g(x + dx)$. Therefore, the two shaded areas in the figure are equal and we conclude

$$f_Y(y)\,dy = f_X(x)\,dx \quad \text{or} \quad f_Y(y) = f_X(x)\frac{dx}{dy} = \frac{f_X(x)}{dy/dx} \equiv \frac{f_X(x)}{g'(x)}. \qquad (1.23)$$

There are three modifications we must make to (1.23) to completely validate it. These anomalies are suggested by Figure 1.20.

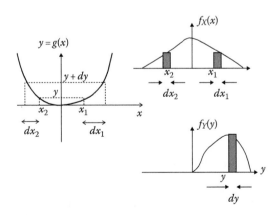

FIGURE 1.20 Changing variables.

1. For a function like $g(x) = x^2$, there is more than one value of x giving rise to the same value y (namely, $x = \pm\sqrt{y}$). So the probability that Y lies between

y and $y + dy$ equals the *sum* of the probabilities that X lies in one of the corresponding intervals $[x_i, x_i + dx_i]$:

$$f_Y(y)dy = f_X(x_1)dx_1 + f_X(x_2)dx_2$$

$$\text{or} \quad f_Y(y) = \sum_{\substack{\text{preimages} \\ \text{of } y}} \frac{f_X(x_i)}{dy/dx|_{x_i}} = \sum_{\substack{\text{preimages} \\ \text{of } y}} \frac{f_X(x_i)}{g'(x_i)}.$$

2. If $g(x)$ is a *decreasing* function of x, then dy and dx will have opposite signs, so to match the areas in the figure, we must state

$$f_Y(y)|dy| = f_X(x_1)|dx_1| + f_X(x_2)|dx_2| + \cdots \quad \text{or} \quad f_Y(y) = \sum_{\substack{\text{preimages} \\ \text{of } y}} \frac{f_X(x_i)}{|g'(x_i)|}. \quad (1.24)$$

3. Finally, since we are going to need the pdf $f_Y(y)$ expressed as a function of y, and not x, each of the preimages x_i has to be expressed in terms of y. For example, suppose that $f_X(x) = e^{-x^2/2}/\sqrt{2\pi}$ and $y = x^2$. Then with $x = \pm\sqrt{y}$ in (1.24) for $y > 0$, we have

$$f_Y(y) = \frac{e^{-x^2/2}/\sqrt{2\pi}}{|2x|}\bigg|_{x=\sqrt{y}} + \frac{e^{-x^2/2}/\sqrt{2\pi}}{|2x|}\bigg|_{x=-\sqrt{y}} = \frac{e^{-y/2}/\sqrt{2\pi}}{|2\sqrt{y}|} + \frac{e^{-y/2}/\sqrt{2\pi}}{|-2\sqrt{y}|} = \frac{e^{-y/2}/\sqrt{2\pi}}{\sqrt{y}}$$

(and $f_Y(y) = 0$ for $y < 0$).*

The trivial relation $Y = X + b$, with b constant, is merely a shift in the variable X, and the relation $Y = aX$, with a constant, is simply a rescaling. Intuitively, we would expect that a simple shift in X would result in a shift of the center of mass μ_X of its pdf graph with no effect on its width σ_X, while rescaling X would rescale both its mean and standard deviation. These observations are special cases of the following formulas for the more general **linear affine** change of variable $Y = aX + b$:

$$\{The\ mean\ of\ Y = aX + b\} = E\{Y\} = \mu_Y = a\mu_X + b. \quad (1.25)$$

$$\{The\ standard\ deviation\ of\ Y = aX + b\} = \sigma_Y = |a|\sigma_X. \quad (1.26)$$

To verify (1.25) and (1.26), one does not need to express the pdf for Y in terms of that for X; they can be proved directly using the linearity property of the expected value and the second-moment identity of Section 1.7.

---

* Don't worry about the ambiguity of $f_Y(0)$; it's not a delta function, so its value at a single point doesn't matter.

SUMMARY: IMPORTANT FACTS ABOUT CHANGE OF VARIABLE

$$pdf \ of \ y = g(x) : f_Y(y) = \sum_{\substack{\text{preimages} \\ \text{of } y}} \frac{f_X(x_i)}{|g'(x_i)|}$$

$$E\{aX + b\} = a\mu_X + b, \ \sigma_{aX+b} = |a|\sigma_X.$$

1.9   THE NORMAL OR GAUSSIAN DISTRIBUTION

When we think of a "typical" probability density function, usually we picture a bell-shaped curve like e^{-x^2} as in Figure 1.21.

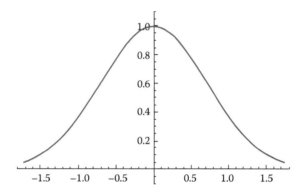

FIGURE 1.21   Bell curve.

There are very good reasons why the bell-shaped curve is so ubiquitous, but we shall not discuss this until Section 1.15. For now, let us contemplate e^{-x^2} as a candidate for a probability density function.

First of all e^{-x^2} must be rescaled so that its integral from $-\infty$ to ∞ is 1. Also, if you look for an antiderivative of e^{-x^2}, you will fail.* It can be shown that e^{-x^2} is not the derivative of any function that can be expressed in closed form. In general, definite integrals of e^{-x^2} have to be calculated numerically using software.

There is a trick, however, for computing the particular definite integral $\int_{-\infty}^{\infty} e^{-x^2} dx$:

$$\text{Let } I = \int_{-\infty}^{\infty} e^{-x^2} dx. \text{ Then } I^2 = \int_{-\infty}^{\infty} e^{-x^2} dx \int_{-\infty}^{\infty} e^{-y^2} dy = \int_{-\infty}^{\infty}\int_{-\infty}^{\infty} e^{-x^2} e^{-y^2} dy \, dx = \iint_{x,y} e^{-(x^2+y^2)} dy \, dx.$$

If we change to polar coordinates (Figure 1.22), the term $x^2 + y^2$ becomes r^2, the element of area $dy \, dx$ becomes $r \, d\theta \, dr$, and the limits become $[0, \infty)$ for r and $[0, 2\pi]$ for θ.

$$I^2 = \int_0^{\infty}\int_0^{2\pi} e^{-r^2} d\theta \, r \, dr = 2\pi \int_0^{\infty} e^{-r^2} d\left(\frac{r^2}{2}\right) = \pi.$$

---

* The branch of analytic function theory known as residue theory will fail also.

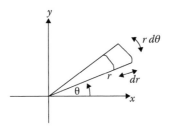

FIGURE 1.22 Integration in polar coordinates.

Therefore, $I = \int_{-\infty}^{\infty} e^{-x^2} dx = \sqrt{\pi}$ and $e^{-x^2}/\sqrt{\pi}$ is a proper probability density function.

Obviously the mean for this pdf is zero, by symmetry. To find its standard deviation, we must take the square root of $\int_{-\infty}^{\infty} x^2 e^{-x^2}/\sqrt{\pi}\, dx$, whose evaluation is no less intractable than the original integral I. But we have another trick. First, we note that we can deduce the integral $\int_{-\infty}^{\infty} e^{-\alpha x^2} dx$ simply by relating it to I:

$$\int_{-\infty}^{\infty} e^{-\alpha x^2} dx = \frac{1}{\sqrt{\alpha}} \int_{-\infty}^{\infty} e^{-(\sqrt{\alpha}x)^2} d(\sqrt{\alpha}x) = \frac{1}{\sqrt{\alpha}} I = \sqrt{\frac{\pi}{\alpha}}.$$

Now, we can get $\int_{-\infty}^{\infty} x^2 e^{-x^2}/\sqrt{\pi}\, dx$ by differentiating $-\int_{-\infty}^{\infty} e^{-\alpha x^2} dx$ with respect to α (yielding $-\int_{-\infty}^{\infty} [-x^2] e^{-\alpha x^2} dx$), and then setting $\alpha = 1$, and finally dividing by $\sqrt{\pi}$. Therefore,

$$\int_{-\infty}^{\infty} x^2 e^{-x^2}/\sqrt{\pi}\, dx = \frac{1}{\sqrt{\pi}} \frac{d}{d\alpha} \left\{ -\sqrt{\frac{\pi}{\alpha}} \right\} \Bigg|_{\alpha=1} = \frac{1}{\sqrt{\pi}} \frac{\sqrt{\pi}}{2} = \frac{1}{2},$$

and the standard deviation is $1/\sqrt{2}$.

In summary, the random variable X with pdf $e^{-x^2}/\sqrt{\pi}$ has mean zero and standard deviation $1/\sqrt{2}$. To get a bell-shaped curve with a different mean μ and standard deviation σ, we shift and rescale; for $Y = aX + b$ to have mean μ and standard deviation σ, then (by Equations 1.25, 1.26 of Section 1.8),

$$\mu = a \cdot 0 + b \ and \sigma = \frac{|a|}{\sqrt{2}}.$$

Thus, $b = \mu$, $a = \sqrt{2}\sigma$, and therefore $Y = \sqrt{2}\sigma X + \mu$ and $X = (Y - \mu)/\sqrt{2}\sigma$. By Equation 1.23 of Section 1.8, then, the derived pdf is $f_Y(y) = \left\{ e^{-\left(\frac{y-\mu}{\sqrt{2}\sigma}\right)^2} / \sqrt{\pi} \right\} (1/\sqrt{2}\sigma)$. In other words, we have shown that

$$f_Y(y) = \frac{e^{-\frac{(y-\mu)^2}{2\sigma^2}}}{\sqrt{2\pi\sigma^2}} \tag{1.27}$$

has mean μ, standard deviation σ, and is bell-shaped. Formula 1.27 is known both as the **normal** or **Gaussian** probability density function, or simply "the normal distribution," and is abbreviated as $N(\mu, \sigma)$.

Figure 1.23 illustrates the normal distribution, displaying the areas (or probabilities) in the critical two-, four-, and six-sigma regions and the tails.*

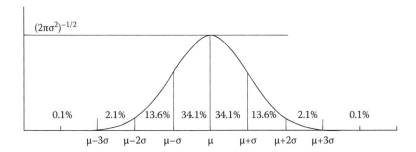

FIGURE 1.23 Normal distribution.

By shifting and rescaling, any definite integral of $N(\mu, \sigma)$ can be expressed as a definite integral of the *standard normal distribution* $N(0, 1)$ (mean zero, standard deviation one). These integrals (the "cumulative normal distribution") have been published to death in statistics books and on the Internet.

Rather than trying to memorize all the details of this important distribution (1.27), you should just keep the following in mind:

1. The denominator is simply the normalizing factor that makes the total area equal to 1.
2. The numerator is the exponential of a *quadratic* in x. In fact *every* function of the form e^{ax^2+bx+c}, with the appropriately scaled denominator, is a normal probability density function as long as $a < 0$.† (If $a \geq 0$, the graph of e^{ax^2+bx+c} houses infinite area and thus is obviously unsuitable as a pdf.)

---

* Before the adoption of the euro, the German ten-mark banknote featured a graph of the normal distribution and a portrait of Gauss.

† To see this, complete the square to express $ax^2 + bx + c$ as $a(x - p)^2 + q$, identify μ as p, identify σ as $1/\sqrt{-2a}$, and incorporate the factor e^q into the normalization. Try it out.

The characteristic function (or inverse Fourier) transform of the normal distribution (1.27) is given by

$$\Phi_X(\omega) = \int_{-\infty}^{\infty} e^{i\omega x} \frac{e^{-\frac{(x-\mu)^2}{2\sigma^2}}}{\sqrt{2\pi\sigma^2}} \, dx = e^{-\left(\sigma^2\omega^2/2\right)+i\mu\omega}.$$

Notice that it is also the exponential of a quadratic! In fact, a theorem by Marcinkiewicz* proves that *any characteristic function of the form* $e^{(\text{polynomial in } \omega)}$ *must be the characteristic function of a normal (Gaussian) distribution* (and therefore that the degree of the polynomial must be 2). (I am taking a few liberties here. If the polynomial is of degree 0 or 1, the characteristic function turns out to be the Fourier transform of a delta function which, with an eye to Figure 1.24, can be interpreted as "a normal with standard deviation zero.")

So an easy way to identify normal distributions is to remember that both the pdfs and the characteristic functions are exponentials of polynomials, and that one needs to specify only the mean and standard deviation to completely determine the distribution.

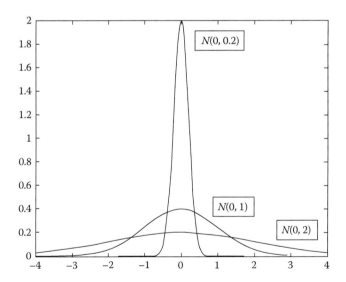

FIGURE 1.24 $N(0, 2)$, $N(0, 1)$, and $N(0, 0.2)$.

---

* Marcinkiewicz, J. 1938. Sur une propriete de la loi de Gauss. *Math. Zeitschrift* 44: 612–618.

SUMMARY: IMPORTANT EQUATIONS INVOLVING
THE NORMAL (GAUSSIAN) DISTRIBUTION

Normal or Gaussian pdf: $N(\mu, \sigma) = f_X(x) = \dfrac{e^{-\frac{(x-\mu)^2}{2\sigma^2}}}{\sqrt{2\pi\sigma^2}}$.

Characteristic function: $\Phi_X(\omega) = \displaystyle\int_{-\infty}^{\infty} e^{i\omega x} e^{-\frac{(x-\mu)^2}{2\sigma^2}} / \sqrt{2\pi\sigma^2}\, dx = e^{-\left(\sigma^2\omega^2/2\right)+i\mu\omega}$.

1.10 MULTIVARIATE STATISTICS

The extension of the techniques of univariate statistics to situations where there are two or more random variables entails a few new wrinkles: the necessity of integrating the pdfs over higher dimensions, the issues of dependence of the variables, and the incorporation of matrix notation. Let's dispose of the first right now. Even in one dimension the normal distribution has to be integrated numerically, so let's not bother with analytic techniques. The following paragraphs provide guidelines for using MATLAB®'s excellent and reliable algorithms for integration in one, two, and three dimensions. Even if you don't know MATLAB, you still should be able to use it to evaluate integrals; just open it (click on the MATLAB icon) and observe how functions are entered into the "@" statements in the following—multiplication, sines, exponentials, division.

1.10.1 ONE-DIMENSIONAL INTEGRALS

To integrate $\displaystyle\int_{1.234}^{5.678} \left[e^{-2x^2} \cos(4x+7) - \dfrac{2x+1}{(x+3)^2(x+1)(x+2)} \right] dx$ in MATLAB, type

    format long (Enter)
    lower = 1.234 (Enter)
    upper = 5.678 (Enter)
    accuracy = 1e–8 (Enter) (Of course you may want more or less accuracy.)

Now, enter the formula for the integrand, *regarding the variable of integration* (x) *as a matrix*. That means inserting the period mark (.) in front of all multiplication signs *, division signs /, and exponentiation carats ^ involving x.

$$\text{integrand} = @(x)\left(\exp(-2*x.^2).*\cos(4*x+7)-(2*x+1).\right.$$
$$\left./((x+3).^2.*(x+1).*(x+2))\right)(\text{Enter})$$

(You'll probably get some typo error message here; retype carefully, counting parentheses and inserting dots.) An error message like

"??? Error using ==> mtimes
Inner matrix dimensions must agree."

means you left out some period marks in front of *, /, or ^. An error message like

"Error: Expression or statement is incorrect--possibly unbalanced (, {, or [."

usually means your parentheses are wrong. *If you change the values of any of the parameters, reenter the integrand=@... statement.*) Now type

Q = quadl(integrand, lower, upper, accuracy) (Enter) (*quadl is quad-ell*)

You should get an answer Q = −0.037041299610442.

1.10.2   TWO-DIMENSIONAL INTEGRALS

To integrate $\int_4^5 \int_6^7 \left(u^2 + uv\right) dv \, du$ in MATLAB, type

format long (Enter)
lowerv = 6 (Enter)
upperv = 7 (Enter)
loweru = 4 (Enter)
upperu = 5 (Enter)
accuracy = 1e−8 (Enter)
integrand = @(u,v)(u.^2 + u.*v) (Enter)
Q = dblquad(integrand, loweru, upperu, lowerv, upperv, accuracy, @quadl) (Enter)
You should get Q = 49.583333333333329.

1.10.3   THREE-DIMENSIONAL INTEGRALS

To integrate $\int_2^3 \int_4^5 \int_6^7 uvw \, dw \, dv \, du$ in MATLAB, type

format long (Enter)
loww = 6 (Enter)
highw = 7 (Enter)
lowv = 4 (Enter)
highv = 5 (Enter)
lowu = 2 (Enter)
highu = 3 (Enter)
accuracy = 1e−8 (Enter)
integrand = @(u,v,w)(u.*v.*w) (Enter)
Q = triplequad(integrand, lowu, highu, lowv, highv, loww, highw, accuracy, @quadl) (Enter)
You should get Q = 73.124999999999986.

When you need to perform numerical integration, be conservative and try out your code with integrands that you can do by hand, to check.

1.11   THE BIVARIATE PROBABILITY DENSITY FUNCTIONS

When there are *two* random variables X and Y to be analyzed, the notion of a probability density function has a natural generalization.

The probability that X lies in the interval $[a, b]$ and Y lies in $[c, d]$ equals the integral

$$\int_a^b \int_c^d f_{XY}(x, y)\, dy\, dx,$$

where $f_{XY}(x, y)$ is known as the **bivariate probability density function**. The graph of f_{XY} is a nonnegative "tent" lying over the x, y plane enclosing unit volume (Figure 1.25).

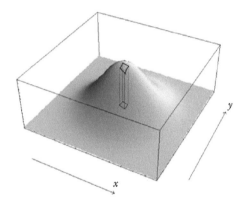

FIGURE 1.25   Bivariate pdf element.

Thus, we say that the probability that X lies between x and $x + dx$ *and* that Y lies between y and $y + dy$ equals $f_{XY}(x, y)\, dx\, dy$.

If we ignore Y and ask for the (total) probability that X lies in the interval $[a, b]$, we get the volume of the "slice" indicated in Figure 1.26. It is found by integrating over the entire range of values for y:

Probability that X lies in the interval $\left[a, b \right] = \int_{x=a}^{b} \int_{y=-\infty}^{\infty} f_{XY}(x, y)\, dy\, dx.$

This has the familiar *uni*variate form $\int_a^b f_X(x)\, dx$, and accordingly we define the

marginal probability density function for X to be $f_X(x) = \int_{y=-\infty}^{\infty} f_{XY}(x, y)\, dy.$

Similarly the marginal density function for Y is $\int_{-\infty}^{\infty} f_{XY}(x, y)\, dx.$ (Visit http://www. uwsp.edu/psych/cw/statmanual/probinter.html to see why this is called the "marginal" probability density function. Believe it or not.)

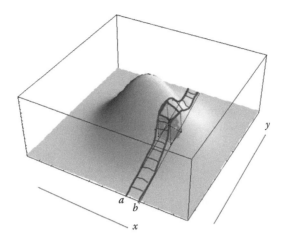

FIGURE 1.26   Marginal probability density.

When there are two random variables the **bivariate expected value of a function** $g(X, Y)$ is the obvious generalization of the univariate expected value:

$$E\{g(X,Y)\} = \int\limits_{-\infty}^{\infty}\int\limits_{-\infty}^{\infty} g(x,y) f_{XY}(x,y) dx\, dy. \tag{1.28}$$

Note that *the* **bivariate mean** *of X*, $\int_{-\infty}^{\infty}\int_{-\infty}^{\infty} x f_{XY}(x,y) dx\, dy$, *is the same as its* **marginal mean**,

$$\int\limits_{-\infty}^{\infty} x f_X(x) dx, \text{ because } \int\limits_{-\infty}^{\infty}\int\limits_{-\infty}^{\infty} x f_{XY}(x,y) dx\, dy = \int\limits_{-\infty}^{\infty} x\left\{\int\limits_{-\infty}^{\infty} f_{XY}(x,y) dy\right\} dx = \int\limits_{-\infty}^{\infty} x f_X(x) dx = \mu_X.$$

Similar reasoning shows that the univariate and marginal means coincide for any *function g(X) of X alone*—for example, its standard deviation or its moments. And the same is true of Y, of course.

An important extension of the *second-moment identity* (Section 1.7) is the following:

$$E\{XY\} = \mu_X\mu_Y + E\{(X - \mu_X)(Y - \mu_Y)\}. \tag{1.29}$$

(To prove it, expand the right-hand side.) $E\{XY\}$ is known as the **correlation** of X and Y.
   Independence: Recall from Section 1.3 that if X and Y are independent, then

[the probability that X lies between x and $x + dx$ and Y lies between y and $y + dy$]

equals

[the probability that X lies between x and $x+dx$]

times

[the probability that Y lies between y and $y+dy$].

In terms of pdfs, this says

> $f_{XY}(x, y) \, dx \, dy$ equals $\{f_X(x) \, dx\}$ times $\{f_Y(y) \, dy\}$, or
> $f_{XY}(x, y) = f_X(x) f_Y(y)$, for independent variables.

Geometrically, independence implies that if we look at two cross sections of the graph of the pdf $f_{XY}(x, y)$ for $y = y_1$ and $y = y_2$, they will have the same shape ($f_X(x)$), but scaled by different factors ($f_Y(y_1)$ and $f_Y(y_2)$). And similarly for constant-x cross sections. Otherwise, the variables are dependent. See Figures 1.27 through 1.29.

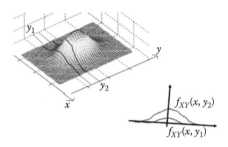

FIGURE 1.27   Independent variables.

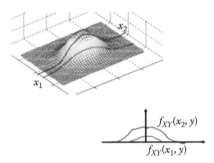

FIGURE 1.28   Independent variables.

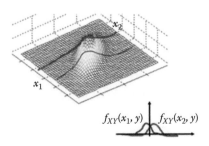

FIGURE 1.29   Dependent variables.

It is important to note that *when the variables X and Y are independent*, the expected value of any function $g(X, Y)$ that can be separated into factors $g_1(X) \, g_2(Y)$ equals the product of the expected values of the factors:

$$E\{g_1(X) \, g_2(Y)\} = E\{g_1(X)\} \, E\{g_2(Y)\} \text{ if } X \text{ and } Y \text{ are independent,}$$

because in the expected value formula $E\{g_1(X) \, g_2(Y)\} = \int_{-\infty}^{\infty} \int_{-\infty}^{\infty} g_1(x) g_2(y)$ $f_{XY}(x, y) \, dx \, dy$, $f_{XY}(x, y)$ splits into $f_X(x) \, f_Y(y)$, and the integral separates:

$$E\{g_1(x) g_2(y)\} = \int_{-\infty}^{\infty} \int_{-\infty}^{\infty} g_1(x) g_2(y) f_{XY}(x, y) \, dx \, dy = \int_{-\infty}^{\infty} \int_{-\infty}^{\infty} g_1(x) g_2(y) f_X(x) f_Y(y) \, dx \, dy$$

$$= \int_{-\infty}^{\infty} g_1(x) f_X(x) \, dx \int_{-\infty}^{\infty} g_2(y) f_Y(y) \, dy = E\{g_1(x)\} E\{g_2(y)\}.$$

Bear in mind that $E\{g_1(X) \, g_2(Y)\} = E\{g_1(X)\} \, E\{g_2(Y)\}$ if X and Y are independent, but $E\{g_1(X) + g_2(Y)\} = E\{g_1(X)\} + E\{g_2(Y)\}$ *even if they are dependent.*

Now if we apply the extended second-moment identity (1.29) to the correlation of *independent* variables, we have

$$E\{XY\} = \mu_X \mu_Y + E\{(X - \mu_X)(Y - \mu_Y)\} \text{(always)}$$

$$= \mu_X \mu_Y \text{ (when } X \text{ and } Y \text{ are independent)}.$$

Therefore, we can take $E\{(X - \mu_X)(Y - \mu_Y)\}$ as a measure of the degree of dependence of X and Y. When it is rescaled by the standard deviations, it is called the **correlation coefficient**:

$$\text{correlation coefficient} = \rho_{XY} = E\{(X - \mu_X)(Y - \mu_Y)\} / (\sigma_X \sigma_Y).$$

Without the rescaling, $E\{(X - \mu_X)(Y - \mu_Y)\}$ is known as the **covariance** of X and Y.

Some algebraic manipulation shows that ρ_{XY} always lies between -1 and 1 (Problem 32).

> **conditional pdf's:** The conditional pdf for X, given that Y takes the value y, is denoted $f_{X|Y}(x|y)$; in other words, the probability that X lies between x and $x + dx$, given that $Y = y$, is $f_{X|Y}(x|y) \, dx$.

Bearing in mind that the statement "$Y = y$" can be interpreted as "Y lies between y and $y + dy$" since dy is infinitesimal, we have by Equation 1.5 of Section 1.3

$$f_{X|Y}(x|y)dx = \frac{f_{XY}(x,y)dx\,dy}{f_Y(y)dy}, \text{ or}$$

$$f_{X|Y}(x|y) = \frac{f_{XY}(x,y)}{f_Y(y)} = \frac{f_{XY}(x,y)}{\int_{-\infty}^{\infty} f_{XY}(x,y)\,dx}. \qquad (1.30)$$

In this context, y is some *given* number—a parameter in Equation 1.30. Thus (1.30) tells us that, as a function of x, the conditional pdf $f_{X|Y}(x|y)$ has the shape of the joint pdf $f_{XY}(x, y)$, for this particular value of y. For a *different* y, the mean of X may shift, its standard deviation may shrink or expand, and its entire pdf may become reshaped. Of course if X and Y are independent, (1.30) reduces to

$$f_{X|Y}(x|y) = \frac{f_X(x)f_Y(y)}{f_Y(y)} = f_X(x),$$

reaffirming that knowledge of Y has no effect on the distribution of X.

The mean and standard deviation of X given Y, which depend on the particular value (y) that Y takes, have to be computed using the conditional pdf $f_{X|Y}(x|y)$:

$$\text{conditional mean of } X = \mu_{X|Y} = \int_{-\infty}^{\infty} x\, f_{X|Y}(x|y)\,dx.$$

$$\text{conditional standard deviation of } X = \sigma_{X|Y}(y) = \sqrt{\int_{-\infty}^{\infty} \left[x - \mu_{X|Y}(y)\right]^2 f_{X|Y}(x|y)\,dx}.$$

Of course analogous equations hold for Y.

ONLINE SOURCES

Correlation coefficient:

McLeod, Ian. "Visualizing Correlations." Wolfram Demonstrations Project. Accessed June 24, 2016. http://demonstrations.wolfram.com/VisualizingCorrelations.

SUMMARY: IMPORTANT EQUATIONS FOR
BIVARIATE RANDOM VARIABLES

Bivariate pdf: probability that X lies between x and $x + dx$ and that Y lies between y and $y + dy$ equals $f_{XY}(x, y)\, dx\, dy$.

Marginal pdf for X is $f_X(x) = \displaystyle\int_{-\infty}^{\infty} f_{XY}(x, y)\, dy$.

Bivariate mean ≡ marginal mean: $\displaystyle\int_{-\infty}^{\infty}\int_{-\infty}^{\infty} x\, f_{XY}(x, y)\, dx\, dy = \int_{-\infty}^{\infty} x\, f_X(x)\, dx = \mu_X$.

Independence: $f_{XY}(x, y) = f_X(x)\, f_Y(y)$,

which implies $E\{g_1(X)\, g_2(Y)\} = E\{g_1(X)\}\, E\{g_2(Y)\}$.
Covariance $= \mathrm{cov}(X, Y) = E\{(X-\mu_X)(Y-\mu_Y)\}$.
Correlation coefficient $= \rho_{XY} = \mathrm{cov}(X, Y)/(\sigma_X \sigma_Y)$, $\quad -1 \le \rho_{XY} \le 1$.
Correlation $= E\{XY\}$.

Extended second-moment identity: $E\{XY\} = \mu_X \mu_Y + E\{(X-\mu_X)(Y-\mu_Y)\}$.

Conditional pdf: $f_{X|Y}(x|y) = \dfrac{f_{XY}(x, y)}{f_Y(y)}$.

Conditional mean: $\mu_{X|Y}(y) = \displaystyle\int_{-\infty}^{\infty} x\, f_{X|Y}(x|y)\, dx$.

Conditional standard deviation: $\sigma_{X|Y}(y) = \sqrt{\displaystyle\int_{-\infty}^{\infty} \left[x - \mu_{X|Y}(y)\right]^2 f_{X|Y}(x|y)\, dx}$.

1.12   THE BIVARIATE GAUSSIAN DISTRIBUTION

If X and Y are univariate Gaussian random variables *and if they are independent*, their bivariate pdf equals

$$f_{XY}(x, y) = \frac{e^{-\frac{(x-\mu_X)^2}{2\sigma_X^2}}\, e^{-\frac{(y-\mu_Y)^2}{2\sigma_Y^2}}}{\sqrt{2\pi\sigma_X^2}\,\sqrt{2\pi\sigma_Y^2}} = \frac{e^{-\frac{1}{2}\left[\frac{(x-\mu_X)^2}{\sigma_X^2} + \frac{(y-\mu_Y)^2}{\sigma_Y^2}\right]}}{2\pi\sigma_X\sigma_Y}. \tag{1.31}$$

Note that f_{XY} has the form $e^{quadratic\ in\ x\ and\ y}$ and that by completing the squares, we can rewrite *any* function of the form $Ke^{Ax^2 + By^2 + Cx + Dy + E}$ in the format of (1.31) (assuming that its integral is unity and A and B are negative).

However, $Ax^2 + By^2 + Cx + Dy + E$ is not the most general second-degree polynomial in x and y, because there is no (xy) term. In fact, the presence of such a term would imply that the variables X and Y are dependent, because the *augmented* expression $e^{Ax^2 + By^2 + Cx + Dy + E + Fxy}$ *cannot* be separated into $f_X(x)$ times $f_Y(y)$.

So we shall take the general definition of the **bivariate Gaussian pdf** to have the $e^{quadratic}$ form with the (xy) term included. A little algebra will convince you that constants $\mu_x, \mu_y, \sigma_x, \sigma_y$, and ρ can be calculated so that every such bivariate Gaussian pdf can be rewritten as

$$
f_{XY}(x,y) = \frac{e^{-\frac{1}{2(1-\rho^2)}\left[\frac{(x-\mu_X)^2}{\sigma_X^2} - 2\rho\frac{(x-\mu_X)(y-\mu_Y)}{\sigma_X \sigma_Y} + \frac{(y-\mu_Y)^2}{\sigma_Y^2}\right]}}{2\pi\sigma_X\sigma_Y\sqrt{1-\rho^2}}.
\tag{1.32}
$$

The new parameter ρ accompanies the (xy) term in the exponential and thus highlights the dependence of X and Y; if $\rho = 0$ the expression reduces to (1.31), the Gaussian pdf for independent variables.

Straightforward (but lengthy) integrations reveal the significance of all the parameters in (1.32):

1. The denominator, of course, normalizes the total integral of f_{XY} to be unity.
2. The marginal pdf for X, $f_X(x) = \int_{-\infty}^{\infty} f_{XY}(x,y)\, dy$, equals $f_X(x) = \dfrac{e^{-(x-\mu_X)^2/2\sigma_X^2}}{\sqrt{2\pi\sigma_X^2}}$,

   which is the univariate normal distribution $N(\mu_X, \sigma_X)$ with mean μ_X and standard deviation σ_X; similarly $f_Y(y) = N(\mu_Y, \sigma_Y)$. Recall from Section 1.11 that the bivariate means and standard deviations coincide with the marginal means and standard deviations.
3. The parameter ρ does not appear in the marginal pdfs; it is the correlation coefficient $\rho = E\{(X-\mu_X)(Y-\mu_Y)\}/(\sigma_X \sigma_Y)$ (Section 1.11). Therefore, ρ always lies between -1 and 1 and is a measure of the dependence of X and Y.
4. The conditional pdf of X given Y, $f_{X|Y}(x|y) = f_{XY}(x,y)/f_Y(y)$, is also a normal distribution with mean and standard deviation

$$
\mu_{X|Y}(y) = \int_{-\infty}^{\infty} x f_{X|Y}(x|y)\, dx = \mu_X + \rho\frac{\sigma_X}{\sigma_Y}(y-\mu_Y)
\tag{1.33}
$$

$$
\sigma_{X|Y}(y) = \sqrt{\int_{-\infty}^{\infty} \left[x - \mu_{X|Y}(y)\right]^2 f_{X|Y}(x|y)\, dx} = \sigma_X\sqrt{1-\rho^2}.
\tag{1.34}
$$

(Keep in mind that y is a given parameter in this context.) So rather than print out $f_{X|Y}(x|y)$ explicitly, we can simply write $N(\mu_{X|Y}(y), \sigma_{X|Y}(y))$. The analogous formulas for the conditional mean and standard deviation of Y given X are

$$
\mu_{Y|X}(x) = \mu_Y + \rho\frac{\sigma_Y}{\sigma_X}(x-\mu_X),
\tag{1.35}
$$

$$
\sigma_{Y|X}(x) = \sigma_Y\sqrt{1-\rho^2}.
\tag{1.36}
$$

Note that we would expect the uncertainty in a probabilistic situation to be reduced if additional information is given, and the reduction in the standard deviations by the factor $\sqrt{1-\rho^2}$ in (1.34, 1.36) reflects this; if X and Y are independent, of course, ρ is zero and there is no improvement, but if $|\rho| \approx 1$, X and Y are strongly correlated, and the "inside information" on Y enables more precise prediction of X.

By the same token, if Y is known to be "running high"—that is, much higher than its expected value—and if it is strongly correlated to X, then we would expect X to "run high" also; that's what Equation 1.33 tells us, for $\rho \approx 1$. If $\rho \approx -1$, then we might (but don't) say X and Y are "anticorrelated" and higher-than-expected values of Y would trigger lower-than-expected values of X.

Some typical graphs of the bivariate normal distribution are displayed in Figures 1.30 through 1.33. Note in particular how the high values of $|\rho|$ indicate a strong relation between X and Y.

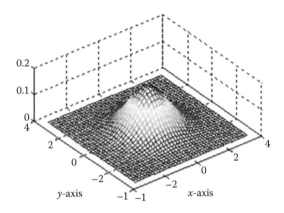

FIGURE 1.30 $\mu_X = \mu_Y = 0$, $\sigma_X = \sigma_Y = 1$, $\rho = 0$.

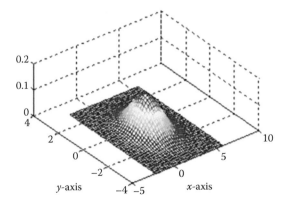

FIGURE 1.31 $\mu_X = 1.8$, $\mu_Y = -0.5$, $\sigma_X = 1.1$, $\sigma_Y = 1$, $\rho = 0$.

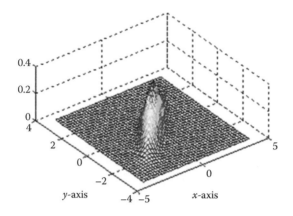

FIGURE 1.32 $\mu_X = \mu_Y = 0$, $\sigma_X = 1.1$, $\sigma_Y = 1$, $\rho = 0.9$.

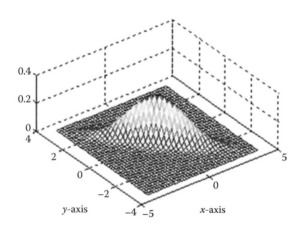

FIGURE 1.33 $\mu_X = \mu_Y = 0$, $\sigma_X = 1.1$, $\sigma_Y = 1$, $\rho = -0.9$.

ONLINE SOURCES

Demonstrations of various aspects of the bivariate Gaussian:

Boucher, Chris. "The Bivariate Normal Distribution." Wolfram Demonstrations Project. Accessed June 24, 2016. http://demonstrations.wolfram.com/TheBivariateNormalDistribution/

Shea, John M.. "Joint Density of Bivariate Gaussian Random Variables." Wolfram Demonstrations Project. Accessed June 24, 2016. http://demonstrations.wolfram.com/JointDensityOfBivariateGaussianRandomVariables/

Boucher, Chris. "The Bivariate Normal and Conditional Distributions." Wolfram Demonstrations Project. Accessed June 24, 2016. http://demonstrations.wolfram.com/TheBivariateNormalAndConditionalDistributions/

$$\text{SUMMARY OF IMPORTANT EQUATIONS}$$
$$\text{FOR THE BIVARIATE GAUSSIAN}$$

Bivariate Gaussian $f_{XY}(x,y) = \dfrac{e^{-\frac{1}{2(1-\rho^2)}\left[\frac{(x-\mu_X)^2}{\sigma_X^2} - 2\rho\frac{(x-\mu_X)(y-\mu_Y)}{\sigma_X\,\sigma_Y} + \frac{(y-\mu_Y)^2}{\sigma_Y^2}\right]}}{2\pi\sigma_X\sigma_Y\sqrt{1-\rho^2}}$

$$[\rho = E\{(X-\mu_X)(Y-\mu_Y)\}/(\sigma_X\sigma_Y)].$$

Marginal of bivariate Gaussian $f_X(x) = \dfrac{e^{-\frac{(x-\mu_X)^2}{2\sigma_X^2}}}{\sqrt{2\pi\sigma_X^2}}.$

Conditional of bivariate Gaussian $f_{X|Y}(x|y) = N\left(\mu_X + \rho\dfrac{\sigma_X}{\sigma_Y}(y-\mu_Y),\, \sigma_X\sqrt{1-\rho^2}\right).$

1.13   SUMS OF RANDOM VARIABLES

Suppose X and Y are independent discrete random variables with probabilities $p(X = x)$, $p(Y = y)$ given in Table 1.1.

TABLE 1.1
Probabilities

x	p(X=x)	y	p(Y=y)
0	0.2	2	0.3
1	0.3	3	0.4
2	0.4	4	0.1
3	0.1	5	0.2

What is the probability that the sum $X + Y$ equals 5? A sum of 5 can arise in four ways: $x = 0$, $y = 5$; $x = 1$, $y = 4$; $x = 2$, $y = 3$; and $x = 3$, $y = 2$. Since X and Y are independent, we can evaluate the probabilities of these pairings by simple multiplication.

$$p(X = 0 \,\&\, Y = 5) = p(X = 0)\,p(Y = 5) = (0.2)(0.2)$$

$$p(X = 1 \,\&\, Y = 4) = p(X = 1)\,p(Y = 4) = (0.3)(0.1)$$

$$p(X = 2 \& Y = 3) = p(X = 2)p(Y = 3) = (0.4)(0.4)$$

$$p(X = 3 \& Y = 2) = p(X = 3)p(Y = 2) = (0.1)(0.3)$$

and the probability that $X + Y$ equals 5 is (0.2) (0.2) + (0.3) (0.1) + (0.4) (0.4) + (0.1) (0.3) or 0.26.

More generally, the probability that $X+Y$ equals z is the product of the probability that $X=x$ times the probability that $Y = z-x$, summed over all the values of x:

$$p(X + Y = z) = \sum_{all\,x} p(X = x)p(Y = z - x).$$

Let's derive the expression for the pdf of the sum of two *continuous* random variables X and Y with joint pdf $f_{XY}(x, y)$. But now for the sake of generality, we do *not* assume them to be independent.

We will start by first deriving the *conditional* pdf for $(X + Y)$, given X. Suppose that X is known to be 2. Then the probability that $Z = X + Y$ lies between, say, 5 and 5.001 is the same as the probability that Y lies between 3 and 3.001, for this value of X. In general, if the value of X is known to be x, then the probability that Z lies between z and $z + dz$ is the same as the probability that Y lies between $(z - x)$ and $(z - x) + dz$ (note: dz, not dy). In terms of conditional pdfs, this says

$$f_{Z|X}(z|x)dz = f_{Y|X}((z - x)|x)\,dz, \text{or } f_{Z|X}(z|x) = f_{Y|X}((z - x)|x). \qquad (1.37)$$

Then to get the pdf for Z, $f_Z(z)$, we use (1.37) and the laws of conditional and marginal probability to derive (think these equalities through one at a time)

$$f_Z(z) = \int_{-\infty}^{\infty} f_{ZX}(z, x)\, dx = \int_{-\infty}^{\infty} f_{Z|X}(z|x)\, f_X(x)\, dx$$

$$= \int_{-\infty}^{\infty} f_{Y|X}((z - x)|x)\, f_X(x)\, dx = \int_{-\infty}^{\infty} f_{XY}(x, z - x)\, dx. \qquad (1.38)$$

Now if X and Y happen to be independent, then the pdf for the sum Z becomes a *convolution*:

$$f_Z(z) = \int_{-\infty}^{\infty} f_X(x)f_Y(z - x)\, dx \text{ when } X,Y \text{ are independent } (Z = X + Y). \qquad (1.39)$$

Equation 1.39 has two very important applications.

First, we consider the case when the independent variables X and Y are Gaussian (normal). Then we claim that the pdf for Z is, also, Gaussian. *Take this for granted for the moment.* We can immediately deduce the formula for $f_Z(z)$, because we can determine its mean and standard deviation:

$$E\{Z\} = \mu_Z = E\{X+Y\} = \mu_X + \mu_Y, \tag{1.40}$$

$$E\left\{(Z-\mu_Z)^2\right\} = \sigma_Z^2 = E\left\{(X-\mu_X+Y-\mu_Y)^2\right\} = \sigma_X^2 + \sigma_Y^2 + 2E\left\{(X-\mu_X)(Y-\mu_Y)\right\}, \tag{1.41}$$

but $E\{(X-\mu_X)(Y-\mu_Y)\} = 0$ for independent variables. Therefore the pdf for Z is $N\left(\mu_X + \mu_Y, \sqrt{\sigma_X^2 + \sigma_Y^2}\right)$.

How did we know that Z is Gaussian? This is an instance of where we use the characteristic function, the inverse Fourier transform of the pdf. Every book on Fourier transforms ever printed proves that the inverse Fourier transform of the convolution of two real functions is the product of the inverse Fourier transforms of each. From (1.39), then, the characteristic function of Z is the product of the characteristic functions of X and Y. But, since X and Y are Gaussian, their characteristic functions are exponentials of quadratics; and when you multiply two such exponentials, you simply add the quadratics—resulting in another quadratic. So the characteristic function for Z is another quadratic exponential, implying that Z, too, is Gaussian. (Wasn't that easy?)

The second application of (1.39) arises when we extend it to the sum of a large number of random variables, $Z = X_1 + X_2 + \ldots + X_n$. For this application, we assume that the X_i are independent and have identical probability density functions: they are independent identically distributed ("iid"). *But we do not assume they are Gaussian.*

Then the pdf of the sum is calculated, according to (1.39), by taking the convolution of the (common) pdf with itself (to get the pdf for X_1+X_2), and then taking the convolution of the result with the original pdf to get the pdf for $X_1+X_2+X_3$, then taking the convolution of *that* result with the original pdf, and so on, n times.

In Figure 1.34, we show the results of such iterated convolutions, starting with a "generic" pdf having no special structure.

The iterated convolutions yield pdfs that become increasing Gaussian in appearance. This is a perfectly general phenomenon: It can be proved that no matter what pdf we start with, by iteratively convolving it with itself, we always approach a Gaussian shape! We shall refer you to the specialized texts to get the precise requirements, the proof, and the nature of the convergence, but for practical purposes we can say that as $n \to \infty$, the pdf $f_Z(z)$ for the sum $Z = X_1 + X_2 + \cdots + X_n$ of n independent, identically distributed random variables approaches the normal distribution with mean $n\mu_X$ and standard deviation $\sqrt{n}\,\sigma_X$ (recall Equations 1.40 and 1.41). This is known as the **Central Limit Theorem.**

We need to interpret the word "approaches" with a little lenience here. If the individual X_i's are discrete and have delta functions in their (common) pdf, then the convolutions will also contain delta functions; they do not converge to the (continuous) normal distributions in the usual mathematical sense. The nature of the convergence in such a case is demonstrated by the coin flip experiment: Here, we take $X_i = +1$ for heads, -1 for tails. In Figure 1.35, the pdfs for the sums of the X_i

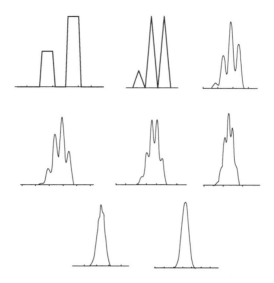

FIGURE 1.34   Iterated convolutions.

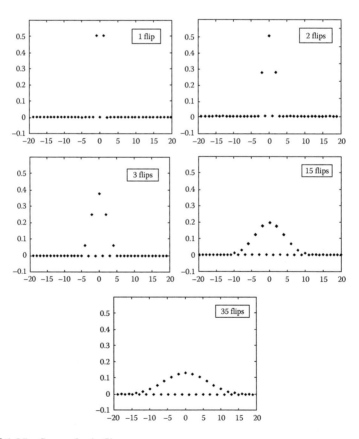

FIGURE 1.35   Sums of coin flips.

are displayed by asterisks indicating the intensities of the delta function components. (The usual "impulse arrows" are omitted.) The fine details of the convergence are explored in Problem 32, but the nature of the compliance with the Central Limit Theorem is clear.

The Central Limit Theorem explains the ubiquitousness of the Gaussian bell-shaped curve in nature. Many random phenomena are the result of a sum of a large number of independent random variables: For example, the noise voltage on a warm resistor results from the thermal agitation of its numerous electrons, and the random fluctuations in a DC current are due to the fact that, at the microscopic level, the "continuous" flow is really a superposition of a huge number of tiny discrete carriers. Therefore, the aggregate phenomena are well modeled by a normal distribution.

A final observation we wish to make about the distribution of sums of random variables is to note the advantage of *averaging* many pieces of data. If we form the sum $Z = X_1 + X_2 + \ldots + X_n$ of n iid random variables, then the calculations in (1.40, 1.41) easily generalize to show $\mu_Z = n\mu_X$ and $\sigma_Z^2 = n\sigma_X^2$. Now divide Z by n, forming the "sample average" $<X> = Z/n$. This is a simple rescaling (i.e., a linear affine change of variable): $<X> = Z/n \ (= aZ + b)$. According to the laws we derived in Section 1.8, then, the mean of $<X>$ is

$$\mu_{\langle X \rangle} = E\left\{\langle X \rangle\right\} = \frac{\mu_Z}{n} = \frac{n\mu_X}{n} = \mu_X$$

(no surprise here!), but its standard deviation is

$$\sigma_{\langle X \rangle} = \frac{\sigma_Z}{n} = \frac{\sqrt{n}\sigma_X}{n} = \frac{\sigma_X}{\sqrt{n}}.$$

We reduce the "spread" in the variable X by the factor $\sqrt{n}$ when we average n samples!

ONLINE SOURCES

Central Limit Theorem:

Gerdy, Jim. "Simulating a Normal Process from Sums of Uniform Distributions." Wolfram Demonstrations Project. Accessed June 24, 2016. http://demonstrations.wolfram.com/SimulatingANormalProcessFromSumsOfUniformDistributions/

Normand, Mark D. and Peleg, Micha. "Central Limit Theorem Applied to Samples of Different Sizes and Ranges." Wolfram Demonstrations Project. Accessed June 24, 2016. http://demonstrations.wolfram.com/CentralLimitTheoremAppliedToSamplesOfDifferentSizesAndRanges/

McLeod, Ian. "Simulated Coin Tossing Experiments and the Law of Large Numbers." Wolfram Demonstrations Project. Accessed June 24, 2016. http://demonstrations.wolfram.com/SimulatedCoinTossingExperimentsAndTheLawOfLargeNumbers/

McLeod, Ian. "Illustrating the Central Limit Theorem with Sums of Uniform and Exponential Random Variables." Wolfram Demonstrations Project. Accessed June 24, 2016. http://demonstrations.wolfram.com/IllustratingTheCentralLimitTheoremWithSumsOfUniformAndExpone/

Watson, David K. "Central Limit Theorem for the Continuous Uniform Distribution." Wolfram Demonstrations Project. Accessed June 24, 2016. http://demonstrations.wolfram.com/CentralLimitTheoremForTheContinuousUniformDistribution/

Brown, Roger J. and Rimmer, Bob. "Generalized Central Limit Theorem." Wolfram Demonstrations Project. Accessed June 24, 2016. http://demonstrations.wolfram.com/GeneralizedCentralLimitTheorem/

SUMMARY OF IMPORTANT EQUATIONS FOR SUMS OF RANDOM VARIABLES

Sum $Z = X + Y$: if X and Y are independent, $f_Z(z) = \int_{-\infty}^{\infty} f_X(x) f_Y(z-x)\, dx$.

If X and Y are independent and Gaussian, $f_Z(z) = N\left(\mu_X + \mu_Y, \sqrt{\sigma_X^2 + \sigma_Y^2}\right)$.

Central Limit Theorem: if $\{X_i\}$ are iid, then

$$Z = \sum_{i=1}^{n} X_i \text{ implies } f_Z(z) \to_{n \to \infty} N\left(n\mu_X, \sqrt{n}\sigma_X\right),$$

$$\langle X \rangle = \frac{1}{n} \sum_{i=1}^{n} X_i \text{ implies } f_{\langle X \rangle}(x) \to_{n \to \infty} N\left(\mu_X, \frac{\sigma_X}{\sqrt{n}}\right).$$

1.14 THE MULTIVARIATE GAUSSIAN

In this section, we are going to rewrite the bivariate normal probability density function in a manner that suggests how it should be extended to apply to n random variables. To do this, we need to review the matrix formulation of quadratic forms.

Observe the expansion of the following matrix product:

$$\begin{bmatrix} x & y \end{bmatrix} \begin{bmatrix} A & B \\ C & D \end{bmatrix} \begin{bmatrix} x \\ y \end{bmatrix} = \begin{bmatrix} x & y \end{bmatrix} \begin{bmatrix} Ax + By \\ Cx + Dy \end{bmatrix} = Ax^2 + (B+C)xy + Dy^2.$$

It expresses a quadratic form in x and y. In fact, the most general quadratic expression $Fx^2 + Gxy + Hy^2$ can be written this way, *using a symmetric matrix*:

$$Fx^2 + Gxy + Hy^2 = \begin{bmatrix} x & y \end{bmatrix} \begin{bmatrix} F & G/2 \\ G/2 & H \end{bmatrix} \begin{bmatrix} x \\ y \end{bmatrix}.$$

More generally, the matrix expression of higher-dimensional quadratic forms proceeds analogously:

$$\begin{bmatrix} x_1 & x_2 & x_3 & \cdots & x_n \end{bmatrix} \begin{bmatrix} a_{11} & a_{12} & a_{13} & \cdots & a_{1n} \\ a_{21} & a_{22} & a_{23} & \cdots & a_{2n} \\ & \cdots & & \ddots & \vdots \\ a_{n1} & a_{n2} & a_{n3} & \cdots & a_{nn} \end{bmatrix} \begin{bmatrix} x_1 \\ x_2 \\ \vdots \\ x_n \end{bmatrix} = \sum_{i=1}^{n} \sum_{j=1}^{n} a_{ij} x_i x_j$$

(and we might as well take the matrix $\{a_{ij}\}$ to be symmetric, since for $i \neq j$ the coefficients only occur in the combination $(a_{ij} + a_{ji})x_i x_j$.

Let's express the exponential in the bivariate Gaussian

$$f_{XY}(x,y) = \frac{e^{-\frac{1}{2(1-\rho^2)}\left[\frac{(x-\mu_X)^2}{\sigma_X^2} - 2\rho\frac{(x-\mu_X)(y-\mu_Y)}{\sigma_X\,\sigma_Y} + \frac{(y-\mu_Y)^2}{\sigma_Y^2}\right]}}{2\pi\sigma_X\sigma_Y\sqrt{1-\rho^2}}$$

as a matrix product. We have

$$\frac{1}{\left(1-\rho^2\right)}\left[\frac{(x-\mu_X)^2}{\sigma_X^2} - 2\rho\frac{(x-\mu_X)(y-\mu_Y)}{\sigma_X\,\sigma_Y} + \frac{(y-\mu_Y)^2}{\sigma_Y^2}\right]$$

$$= \begin{bmatrix} x-\mu_X & y-\mu_Y \end{bmatrix} \begin{bmatrix} \dfrac{1}{\sigma_X^2\left(1-\rho^2\right)} & \dfrac{-\rho}{\sigma_X\sigma_Y\left(1-\rho^2\right)} \\[2mm] \dfrac{-\rho}{\sigma_X\sigma_Y\left(1-\rho^2\right)} & \dfrac{1}{\sigma_X^2\left(1-\rho^2\right)} \end{bmatrix} \begin{bmatrix} x-\mu_X \\ y-\mu_Y \end{bmatrix}.$$

Now the inverse of this matrix is easily calculated, and it reveals that the terms therein are quite familiar:

$$\begin{bmatrix} \dfrac{1}{\sigma_X^2\left(1-\rho^2\right)} & \dfrac{-\rho}{\sigma_X\sigma_Y\left(1-\rho^2\right)} \\[2mm] \dfrac{-\rho}{\sigma_X\sigma_Y\left(1-\rho^2\right)} & \dfrac{1}{\sigma_Y^2\left(1-\rho^2\right)} \end{bmatrix}^{-1} = \begin{bmatrix} \sigma_X^2 & \rho\sigma_X\sigma_Y \\ \rho\sigma_X\sigma_Y & \sigma_Y^2 \end{bmatrix}$$

$$= \begin{bmatrix} E\left\{(X-\mu_X)^2\right\} & E\left\{(X-\mu_X)(Y-\mu_Y)\right\} \\ E\left\{(X-\mu_X)(Y-\mu_Y)\right\} & E\left\{(Y-\mu_Y)^2\right\} \end{bmatrix}.$$

The latter matrix is known as the *covariance* matrix "cov." For n variables its i,jth entry is $E\{(X_i - \mu_i)(X_j - \mu_j)\}$.

Therefore, the matrix appearing in the quadratic form for the bivariate Gaussian is the inverse of the covariance matrix, and we can write this pdf as

$$f_{XY}(x,y) = \frac{e^{-\frac{1}{2}\begin{bmatrix} x-\mu_X & y-\mu_Y \end{bmatrix}[cov]^{-1}\begin{bmatrix} x-\mu_X \\ y-\mu_Y \end{bmatrix}}}{2\pi\sigma_X\sigma_Y\sqrt{1-\rho^2}}. \qquad (1.42)$$

Furthermore, the determinant of the covariance matrix is $\sigma_X^2\sigma_Y^2\left(1-\rho^2\right)$, so the bivariate Gaussian pdf is neatly expressed by

$$f_{XY}(x,y) = \frac{e^{-\frac{1}{2}\begin{bmatrix} x-\mu_X & y-\mu_Y \end{bmatrix}[cov]^{-1}\begin{bmatrix} x-\mu_X \\ y-\mu_Y \end{bmatrix}}}{2\pi\sqrt{\det(cov)}}. \qquad (1.43)$$

Now, compare this with the univariate Gaussian pdf for which the covariance matrix is $E\{(X - \mu_x)^2\} = \sigma_x^2$:

$$f_X(x) = \frac{e^{-\frac{1}{2}\frac{(x-\mu)^2}{\sigma^2}}}{\sqrt{2\pi\sigma^2}} = \frac{e^{-\frac{1}{2}[x-\mu_X][cov]^{-1}[x-\mu_X]}}{\sqrt{2\pi}\sqrt{\det(cov)}}. \tag{1.44}$$

Comparison of formulas (1.43) and (1.44) tells us how to generalize the Gaussian probability density function to n variables written in row-vector form as $\mathbf{X} = [X_1, X_2, \ldots, X_n]$ with means $\mu = [\mu_1, \mu_2, \ldots, \mu_n]$ and covariance matrix $\mathbf{cov}$; namely,

$$f_\mathbf{X}(\mathbf{x}) = \frac{e^{-\frac{1}{2}[\mathbf{x}-\mu][\mathbf{cov}]^{-1}[\mathbf{x}-\mu]^T}}{(2\pi)^{n/2}\sqrt{\det(\mathbf{cov})}}. \tag{1.45}$$

1.15   THE IMPORTANCE OF THE NORMAL DISTRIBUTION

As you study more advanced texts and research articles on random processes, it will occur to you that the vast majority of them presume that the underlying distributions are normal (Gaussian). In fact, every deviation from "normality" usually warrants a stand-alone publication. So we close Chapter 1 with a "top-ten" list of the advantageous properties of this remarkable distribution.

Ode to the Normal Distribution

How do I love thee? Let me count the ways.

1. The Gaussian's bell-shaped curve is ubiquitous in nature.
2. The Gaussian's pdf is characterized by two parameters: mean and standard deviation. (But this is true of many other distributions as well.*)
3. Any linear affine transformation $(aX + b)$ of Gaussian variables is Gaussian. (So we only need to tabulate a standardized version using $[X-\mu]/\sigma$.)
4. Joint Gaussians have Gaussian marginals.
5. Joint Gaussians have Gaussian conditionals.
6. Independence is equivalent to zero correlation. (This is true of some other distributions as well.†)
7. Any sum of Gaussian variables is Gaussian.
8. The Central Limit Theorem says that sums of large numbers of independent identically distributed random variables have distributions that "approach" Gaussian, *even if the individual variables are not Gaussian.*

---

* Indeed, the Poisson and exponential distributions are characterized by only one parameter.
† Lancaster, H.O. 1959. Zero correlation and independence. *Aus. J. Stat.* 1: 53–56.

9. The Gaussian's characteristic (moment generating) function is Gaussian.
10. *Any* characteristic function whose logarithm is a polynomial is the characteristic of a Gaussian (and the polynomial is a quadratic).
    Well, what is there not to like about the Gaussian?

Answer: you can't integrate e^{-x^2} in closed form.

EXERCISES

SECTION 1.1

1. Generalize the game show problem. If there is a prize behind one of four doors, by how much does your probability of winning change if you switch choices after a losing door is opened? What about n doors?

SECTION 1.3

2. Analyze the game show problem in Section 1 using Bayes' theorem.
3. Box 1 contains 100 resistors of which 10 percent are defective (they do not conduct). Box 2 contains 200 resistors of which 5 percent are defective. Two resistors are picked from a randomly selected box.
  a. If a circuit branch is formed by connecting the resistors in parallel (Figure 1.36a), what is the probability that the branch will conduct?
  b. If the resistors are connected in series (Figure 1.36b), what is the probability that the branch will conduct?
  c. If the parallel branch does not conduct, what is the probability that the resistors come from box 1?
  d. If the series branch does not conduct, what is the probability that the resistors come from box 1?

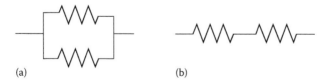

(a)   (b)

FIGURE 1.36   Resistor configurations: (a) parallel (b) series.

SECTION 1.4

4. If there are 30 students in a class, what is the probability that two or more will have the same last 3 digits in their social security numbers (in the same order)? Assume that all digits 0 to 9 are equally likely. (Hint: Look up the "birthday problem" in your basic statistics book or on the Internet.) (Another hint: Asking your computer to evaluate large factorials may result in overflow. Explore the MATLAB™ function "prod" instead.)

5. In the previous problem, what is the smallest number n such that the probability that two or more students will have the same last 3 digits in their social security numbers (in the same order) exceeds 0.5?

6. What is the reasoning behind the ranking of hands in poker? If five playing cards are drawn at random from a 52-card deck, what are the probabilities of the various poker hands that can arise (a single pair, two pairs, three of a kind, a flush, a straight, a full house, four of a kind, a straight flush, a royal straight flush)?

SECTION 1.5

7. A resistor value is measured using a digital multimeter that displays only two digits after rounding off (e.g., instead of 18.56 or 19.44, it displays 19). Let the displayed value be denoted by X. Assuming that the true resistance is uniformly distributed in the range $[18, 22]$ ohms, find the probability density function for X.

8. Suppose that X is a random number uniformly distributed in the interval $[0, 5]$. X is measured and its value is rounded to 3 decimals, resulting in the value 4.377; in other words, $4.3765 \leq X < 4.3775$.
   a. What is the conditional pdf of X, given that its rounded value is 4.377?
   b. What is its conditional mean?
   c. What is its conditional standard deviation?
   d. Some experimentalists would report the result of the measurement as $X = 4.377 \pm 0.0005$; others might quote $X = 4.377 \pm 0.00029$. What is the difference in interpretation?

9. Generalize Problem 8: If X has the pdf $f_X(x)$ and its value, after rounding to d decimal places, is known to be X_{round}, give formulas for its conditional pdf, mean, and standard deviation. Restate your result for rounding to d *binary* places.

10. Suppose that X is a random number uniformly distributed in the interval $[0, 5]$. X is measured and its value is *truncated* to 3 decimals, resulting in the value 4.377; in other words, $4.377 \leq X < 4.378$. What is the conditional pdf of X, given that its truncated value is 4.377? What are the conditional mean and standard deviation?

11. Generalize Problem 10: If X has the pdf $f_X(x)$ and its value, after truncating to d decimal places, is known to be X_{trunc}, what are its conditional pdf, mean, and standard deviation? Restate your result for truncating to d *binary* places.

12. Explain how you can introduce an *offset* to convert a truncating voltmeter (Problem 10) to a rounding voltmeter (Problem 8).

SECTION 1.6

13. How is it possible that an event with probability 1 can fail to occur? (Hint: Consider the random variable X described by the pdf in Figure 1.9; compare the events
   A: $-0.5 \leq X \leq 0.5$, and B: $-0.5 < X < 0.5$.)

SECTION 1.7

14. Confirm the cited values of the means and standard deviations in Figure 1.18.
15. The standard deviation σ provides a measure of how closely a random variable X can be expected to stay near its mean μ. (See Figure 1.18.) **Chebyshev's inequality** states, in fact, that the probability of $|X - \mu|$ exceeding any positive number M falls off faster than $(\sigma/M)^2$: $\mathrm{Prob}\{|X - \mu| > M\} \le (\sigma/M)^2$. Derive the inequality by writing the integral for σ^2 (from 1.20) and restricting the integral to values of $|X - \mu|$ exceeding M.
    Online source: Demonstrations with the Chebyshev inequality can be found at http://demonstrations.wolfram.com/ChebyshevsInequality/.
16. Consult a table of Fourier integrals to find the characteristic function for the **exponential distribution** $\lambda e^{-\lambda t}$, $t \ge 0$ (and 0 for $t < 0$). Use Equations 1.22 of Section 1.7 to find its mean and standard deviation.
17. Consult a table of Fourier integrals to find the characteristic function for the **Erlang probability distribution functions**

$$f_{Erlang}(t) = \begin{cases} 0, & t < 0, \\ \dfrac{\lambda(\lambda t)^{n-1} e^{-\lambda t}}{(n-1)!}, & t \ge 0, \end{cases}$$

where $n = 1, 2, 3, \ldots$ and λ is a parameter. Use Equations 1.22 of Section 1.7 to find the means and standard deviations.

SECTION 1.8

18. Prove (1.25, 1.26) in Section 1.8.
19. If θ is uniformly distributed between 0 and 2π, what are the pdfs of $\sin\theta$, $\cos\theta$, and $\tan\theta$?
20. If X is uniformly distributed in $(0,1)$, what is the pdf of $Y = -\ln x$?
21. In an LC circuit (Figure 1.37) the natural frequency of oscillation is given by the expression $\omega_0 = 1/\sqrt{LC}$. Assume that the capacitance $C = 10\mu F$ but that the value of the inductance L is random, uniformly distributed in the range $9mH \pm 10\%$.
    a. Find the probability density function of the natural frequency.
    b. Find the mean for this density function. Is it the same as the frequency corresponding to the mean inductance $L = 9mH$?

FIGURE 1.37 *LC* circuit.

22. A saturated amplifier "clips" an input signal X in accordance with

$$Y = g(X) = \begin{cases} -1, & X \le -1 \\ X, & -1 < X < 1. \\ +1, & X \ge 1 \end{cases}$$

Describe the pdf for the output Y in the terms of the pdf for X.

23. a. Evaluate the mean, standard deviation, and second moment of the variable Y whose pdf is depicted in the first graph of Figure 1.38a by integration.

   b. Confirm your answers by expressing Y as a scaled-and-shifted version of the random variable X depicted in the centered graph of Figure 1.18, and applying the "derived distributions" formulas (1.25, 1.26) of Section 1.8.

   c. Consider a discrete version of this problem; suppose Z takes the evenly spaced values {1.0, 1.5, 2.0, 2.5, 3.0, 3.5, 4.0) with equal probabilities (see Figure 1.38b). Work out the mean and standard deviation of Z and compare them with those of Y.

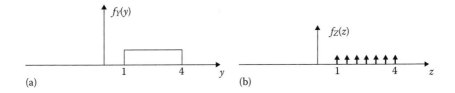

FIGURE 1.38 (a) Continuous variable and (b) discretized variable.

Section 1.9

24. Assume X is a Gaussian random variable with mean 1 and standard deviation 2. Answer the following questions to within 0.01.

   a. What is the probability that the value of X lies between 3 and 4?

   b. If X is known to be greater than 3.5, what is the probability that the value of X lies between 3 and 4?

25. If X is the random variable described in Problem 24 and $Y = X^2 - X + 1$, what is the expected value of Y?

26. If X is a Gaussian random variable with mean zero and standard deviation 1, and $Y = X^3$ and $Z = X^4$, what are the pdfs $f_Y(y)$ and $f_Z(z)$? Are they Gaussian?

27. When X is a standardized Gaussian random variable, $N(0,1)$, and the change of variables $y = e^x$ is introduced, the resulting distribution for Y is called **lognormal**. Derive the pdf for the lognormal distribution.

Online source: Demonstrations with the lognormal distribution can be found at Boucher, Chris. "The Log Normal Distribution." Wolfram Demonstrations Project. Accessed June 24, 2016. http://demonstrations.wolfram.com/TheLog NormalDistribution/

28. If X is normally distributed $N(\mu, \sigma)$, then the pdf of $[X - \mu]/\sigma$ is $N(0, 1)$. Is it true for *any* distribution that $[X - \mu]/\sigma$ has mean zero and standard deviation unity? (Demonstrate.)

29. If the pdf of a random variable X is proportional to e^{-3x^2+2x}, show that X is normal. What are its mean and standard deviation?

30. X is uniformly distributed between -1 and 1. Y has the same mean and standard deviation as X, but is normally distributed. Compare the probabilities that $-0.5 < X < 0.5$ and $-0.5 < Y < 0.5$.

31. We are trying to detect the presence of a 1 V DC (constant) voltage using an imperfect voltmeter. If the voltage is not present, the voltmeter output is normally distributed, $N(0, 0.1)$. If the voltage is present, the reading is distributed as $N(1, 0.1)$. To decide if the voltage is present, we compare the receiver output to 0.5v, and if the output is larger than 0.5, we conclude that the signal is present. Find the probability of the following two types of errors: (a) a miss, when we say the signal is absent but it is present, and (b) a false alarm, when we say that the signal is present but it is not.

SECTION 1.11

32. Show that the correlation coefficient always lies between -1 and 1.

33. We place at random n points in the interval $(0, 1)$ and we denote by X and Y the distance from the origin to the first and last point, respectively. Find $f_X(x)$, $f_Y(y)$, and $f_{XY}(x, y)$. (Hint: It may be easier to find the cumulative distribution functions first.)

34. Suppose X_1 and X_2 are independent identically distributed random variables with $f(x)$ (continuous) as their common pdf. Order them, so that $Y_1 = \min(X_1, X_2)$ and $Y_2 = \max(X_1, X_2)$. Show that the joint distribution function for Y_1 and Y_2 is $f_{Y_1 Y_2}(y_1, y_2) = 2f(y_1)f(y_2)$ for $y_1 < y_2$, and zero otherwise. Generalize: If $\{Y_1, Y_2, ..., Y_n\}$ is the reordering of the iid random variables $\{X_1, X_2, ..., X_n\}$ so that $Y_1 < Y_2 < \cdots < Y_n$, then $f_{Y_1 Y_2 \cdots Y_n}(y_1, y_2, ..., y_n) = n! f(y_1)f(y_2) \cdots f(y_n)$ if $y_1 < y_2 < \cdots < y_n$ and zero otherwise. (Discuss replacing $<$ with $\leq$.)

35. Reconsider Problem 34. Define Z_1 to be the *rank* of X_1, in the sense that $Z_1 = 1$ if $X_1 < X_2$ and $Z_1 = 2$ if $X_2 < X_1$. (Ignore the zero-probability event of a tie.) Similarly let Z_2 equal the rank of X_2. So the possibilities for $\{Z_1, Z_2\}$ are $\{1, 2\}$ and $\{2, 1\}$. What are the probabilities of these two outcomes? Generalize: Let $\{Z_1, Z_2, ..., Z_n\}$ be the ranks of iid $\{X_1, X_2, ..., X_n\}$. What is the probability distribution for $\{Z_1, Z_2, ..., Z_n\}$?

SECTION 1.12

36. Let X and Y be joint Gaussian random variables with $\mu_X = 1$, $\sigma_X = 1$, $\mu_Y = -1$, $\sigma_Y = 2$, $\rho = 0.5$.
   a. What is the probability that $-1 < X < 1$ and $-2 < Y < 2$?
   b. What is the probability that $-2 < Y < 2$?
   c. If you measure X and get $X = 3$, then what is the probability that $-2 < Y < 2$?

Section 1.13

37. Flip a coin: $X = 1$ if you get heads, $X = 2$ if you get tails. If you average the outcome of 100 flips, estimate the probability that $1.3 < \langle X \rangle < 1.6$. (Use the Central Limit Theorem.)

38. Suppose you flip a coin 100 times. Use the Central Limit Theorem to estimate the probability that the number of heads exceeds 60.

39. Use computer simulation to demonstrate that repeated convolutions of an arbitrary nonnegative function converge to (a scaled replica of) the Gaussian. Most software packages have a "convolution" subroutine, but feel free to use the MATLAB™ code below. Be sure to experiment with a variety of exotic (nonnegative) functions by suitably modifying the first statement.

```
A = [zeros(1, 10) ones(1, 5) zeros(1, 5) 2*ones(1, 5)
zeros(1, 10)];
A=A/(sum(A));
disp('Press ENTER to see the first 15 convolutions of the
profile.')
pause;
plot(A);
B = A;
C = conv(A,B);
pause;
plot(C);
pause;
for jj=1:15;
   B = C;
   C = conv(A,B);
   plot(C);
   pause;
end;
```

40. Work out the formulas corresponding to (1.38 through 1.41, Section 1.13) for the *difference X-Y*.

41. Let X and Y be Gaussian random variables with correlation coefficient ρ. Why are the random variables $U = aX + bY$ and $V = cX + dY$ also Gaussian? What conditions on a, b, c, and d will make U and V independent?

42. The coin-flipping experiment depicted in Figure 1.35 of Section 1.13 can be replicated using the following MATLAB code:

```
A=zeros(1,41);
A(20)=.5;A(22)=.5; X=linspace(-20, 20, 41);
disp('Press ENTER to see the first 15 convolutions of the
profile.')
pause;
plot(X,A,'k*');
axis([-20 20 -.1 .6])
B = A;
C = conv(A,B);
C=C(1,21:61);
```

```
pause;
plot(X,C,'k*');
axis([-20 20 -.1 .6])
pause;
for jj=1:50;
   B = C;
   C = conv(A,B);C=C(1,21:61);
   plot(X,C,'k*');
   axis([-20 20 -.1 .6])
   pause;
end;
```

Why are there delta functions of intensity zero interspersed between the nonzero components in Figure 1.35?

2 Random Processes

We can usually recognize a random process when we see one. But if we hope to perform mathematical analysis on random processes, we have to postulate some specific structure that we can work with. So let's consider a few examples and see what they have in common.

2.1 EXAMPLES OF RANDOM PROCESSES

Example 2.1

The random process everyone would like to be able to predict is the fluctuation of the share price of a particular stock! Honest stock traders' skills in predicting this behavior rely on knowledge of the market in general and a study of the behavior of the stock in the past. Dishonest traders use "inside information" to concoct superior models, and the public at large is righteously infuriated when they are exposed. Figure 2.1 displays histories for three typical stocks.

FIGURE 2.1   Stock market samples.

Example 2.2

The temperature at a particular weather station has an overall 12-month pattern, but the fluctuations are random. In past years, forecasters relied heavily on records of previous weather performances, but now computational physics models have improved prediction tremendously. Of course surprises still occur. Figure 2.2 shows a temperature graph spanning 3 years.

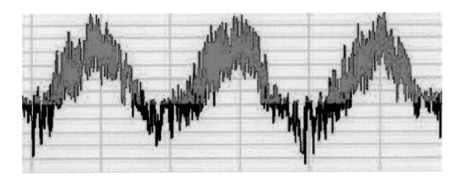

FIGURE 2.2   Three-year temperature chart.

Example 2.3

The thermal motions of electrons in a resistor give rise to small random voltage fluctuations at its terminals. The statistical properties of this thermal noise were analyzed in 1928 by Johnson and Nyquist[*,†] using thermodynamic arguments. A sample of *Johnson noise* is displayed in Figure 2.3.

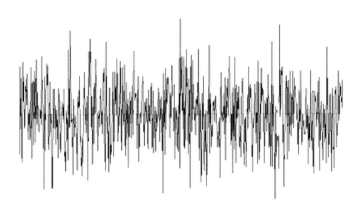

FIGURE 2.3   Johnson noise.

---

[*] Johnson, J. 1928. Thermal agitation of electricity in conductors. *Phys. Rev.* 32: 97.
[†] Nyquist, H. 1928. Thermal agitation of electric charge in conductors. *Phys. Rev.* 32: 110.

Example 2.4

The emission of electrons across a junction proceeds by a random process that can be modeled very well using Poisson statistics.* See Figure 2.4 for a sample of this *shot noise*.

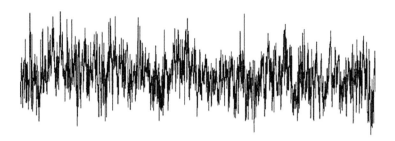

FIGURE 2.4   Shot noise.

Example 2.5

Voltage levels in semiconductor devices sometimes jump spontaneously between discrete levels. Played through a loudspeaker, the effect suggests the name *popcorn noise* (Figure 2.5).

FIGURE 2.5   Popcorn noise.

Example 2.6

As we shall see in Chapter 4, we can *simulate* a random process with speci-fied statistical properties using a linear filter with feedback driven by a random number generator. The so-called Yule–Walker equations enable the computa-tion of an autoregressive-moving-average (ARMA) random process with known autocorrelations. Figure 2.6 is a simulation of an ARMA process, executed using MATLAB®.

---

* Campbell, N. 1909. The study of discontinuous phenomena. *Proc. Camb. Philos. Soc.* 15: 117–136.

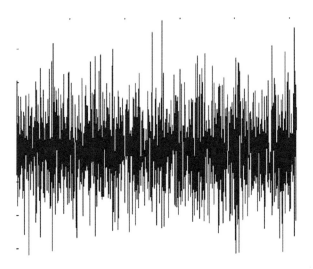

FIGURE 2.6   ARMA simulation.

Example 2.7

Starting at $t = 0$, a coin is flipped every second and we define $X(n)$ to be zero if the flip at the nth second is tails, and one if it is heads. A sample of the process with outcome THTHHHTHTH… is depicted in Figure 2.7. This is called a "Bernoulli process."*

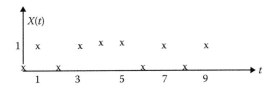

FIGURE 2.7   Bernoulli process.

Example 2.8

Suppose we have a DC power supply whose (constant) output voltage is a random variable uniformly distributed between −1 and 1. (For example, one night a mischief-maker breaks into the lab and randomly resets the "output" dial.) Then its output–voltage timeline is a random process with very simple statistics; see Figure 2.8.

---

* Bernoulli, J. 1713. Ars Conjectandi, Opus Posthumum. Accedit Tractatus de Seriebus infinitis, et Epistola Gallice scripta de ludo Pilae recticularis. *Impensis Thurnisiorum, Fratrum.* Basel, Switzerland.

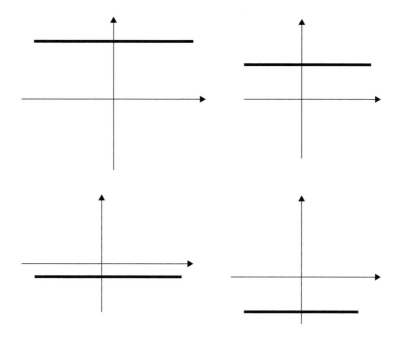

FIGURE 2.8   Random settings for a DC power supply.

Example 2.9

Similarly, if the saboteur targets our lab's sinusoidal signal generator $X(t) = A\sin\omega t$ and resets the amplitude A to a random variable uniformly distributed between 0 and 1, the five graphs in Figure 2.9 display some possible outcomes we might get.

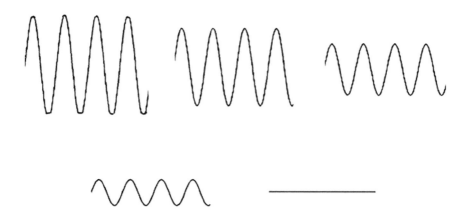

FIGURE 2.9   Random settings for an AC power supply.

Example 2.10

If the saboteur randomly resets the *phase* control on a sinusoidal signal generator $X(t) = \sin(\omega t + \Phi)$, so that Φ becomes, say, a random variable uniformly distributed between 0 and 2π, samples of the possible outcomes would look like those in Figure 2.10.

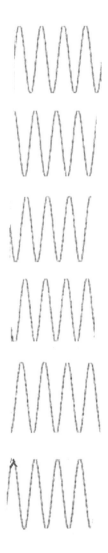

FIGURE 2.10   AC power supply voltages with random phase.

You will have noticed significant differences between these examples. The last three are "idealized" or "contrived"; they are much simpler in the sense that if we measure one or two values of $X(t)$, we could perfectly predict every other value.

But for the first six random processes, no *sub*set of the data from a sample function from any of them would completely remove the uncertainty in the remaining data. Also the first two processes have no theoretical underpinning. There is no way we can *derive* the statistical properties of the stock market or the weather, and we only have the raw data to guide us in making an analysis, whereas we ought to be able to deduce features of the behavior of Johnson and shot noise, and ARMA and Bernoulli processes from the underlying assumptions governing them.

Chapter 3 will focus on the mathematical methods for extracting statistical properties from raw data alone, while Chapter 4 will explore the analysis of many of the theoretical statistical models that have been devised to describe specific random processes. The final chapters detail how the results are used to make informed predictions about the process.

2.2 THE MATHEMATICAL CHARACTERIZATION OF RANDOM PROCESSES

With the examples of Section 2.1 in mind, what would be the reasonable assumptions we can make on the mathematical structure defining a random process?

First of all, $X(t)$ is a function of time. The "randomness" of X is reflected in the fact that at any particular time τ, the value of $X(\tau)$ is a random variable. So if we are to perform any kind of mathematical analysis on it, we must presume that the possible values of $X(\tau)$ are described by a probability density function (pdf) $f_{X(\tau)}(x)$; the probability that the value of X at time τ lies between a and b is $\int_{x=a}^{b} f_{X(\tau)}(x)dx$ (Section 1.5).

Let's be very explicit here. We freeze the value of the time at $t = \tau$; the values of the function $X(t)$ for values of t other than τ are irrelevant for the moment. The probability density function $f_{X(\tau)}(x)$ is being integrated over the variable x, which designates the range of possible values for X at this time. This range of values may be continuous, like the temperature or voltage, or discrete, like stock market prices. It may even be a single value, if X happens to be nonrandom or deterministic. We can accommodate all these possibilities using delta functions in the density function $f_{X(\tau)}(x)$ (Section 1.5). See Figure 2.11.

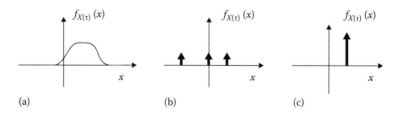

FIGURE 2.11 (a) Continuous, (b) discrete, and (c) deterministic (no uncertainty) pdf's for a random process.

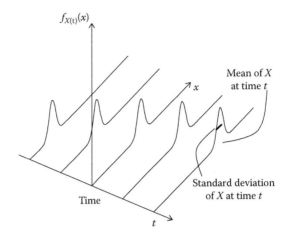

FIGURE 2.12   pdf's for $X(t)$ at different times.

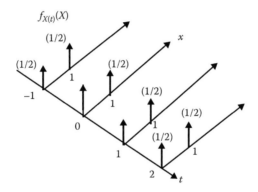

FIGURE 2.13 $f_{X(t)}(x)$ for the Bernoulli process.

Now let's widen our focus and contemplate $X(t)$ for *all* times. Thus, we interpret $f_{X(t)}(x)$ as a *family* of pdfs* describing the range of possible values for X at each time t (Figure 2.12). The figure is really too "pretty"; at this point, there is no reason to presume that the pdf's for different times are related in any way.

The *times* at which X is specified may be either continuous or discrete. (Of course, in practical situations, we almost always measure *samples* of a continuous function, rendering time as discrete.) And we have seen that the *values* that X may take may also be continuous or discrete; all four combinations (discrete t and continuous X, etc.) are possible. Study Figure 2.13 and be sure that you understand how it depicts the pdf for the Bernoulli process described in Section 2.1.

Now we impose some more structure on the model. In most situations, the values that X takes at different times are not independent. If we measure X and observe that X took the value say, 1.07, at time t_1, then we would expect that for times t near t_1 X will probably take a value close to 1.07. That is, the *conditional* pdf for $X(t)$, *given that at time t_1*

---

* *probability density functions.*

X took the value x_1, may have a different shape from the *a priori* pdf $f_{X(t)}(x)$ (Section 1.3). So in addition to postulating that there are pdf's $f_{X(t)}(x)$ for every time t, we must postulate that there are also conditional pdf's for the likelihood that $X(t_2)$ takes the value x_2, given that $X(t_1)$ took the value x_1. The notation for such a conditional pdf is formidable:

$$f_{X(t_2)|X(t_1)}\left(x_2 \,|\, x_1\right). \tag{2.1}$$

For the random DC power supply, $f_{X(t_2)|X(t_1)}\left(x_2 \,|\, x_1\right)$ is trivial because if the voltage is x_1 at one time, it is always x_1: $f_{X(t_2)|X(t_1)}\left(x_2 \,|\, x_1\right) = \delta\left(x_2 - x_1\right)$. And if I had the knowledge of $f_{X(t_2)|X(t_1)}\left(x_2 \,|\, x_1\right)$ for the stock market, I could stop writing textbooks.

Of course the conditional pdf (2.1) is related to the *joint* pdf for $X(t_1)$ and $X(t_2)$ by the usual formula (1.30) in Section 1.11:

$$f_{X(t_1),X(t_2)}\left(x_1, x_2\right) = f_{X(t_1)}\left(x_1\right) f_{X(t_2)|X(t_1)}\left(x_2 \,|\, x_1\right). \tag{2.2}$$

Reasoning along the same line, if we measure X at *two* different times t_1 and t_2, then its pdf at any third time t_3 will likely be affected, giving rise to the conditional pdf

$$f_{X(t_3)|X(t_1)\&X(t_2)}\left(x_3 \,|\, x_1, x_2\right),$$

which is related to joint pdf's via

$$f_{X(t_1),X(t_2),X(t_3)}\left(x_1, x_2, x_3\right) = f_{X(t_1)}\left(x_1\right) f_{X(t_2)|X(t_1)}\left(x_2 \,|\, x_1\right) f_{X(t_3)|X(t_1)\&X(t_2)}\left(x_3 \,|\, x_1, x_2\right)$$

And don't forget the marginal probability relation,

$$f_{X(t_1)}\left(x_1\right) = \int_{-\infty}^{\infty} f_{X(t_1),X(t_2)}\left(x_1, x_2\right) dx_2.$$

So we begin to appreciate the complexity behind the probabilistic description of a random process. Theoretically speaking, there is a mind boggling infinity upon infinity of conditional and joint pdf's involved. And they must all be consistent with the rules for marginal and conditional pdf's.

There are two encouraging notes in this sea of complexity. Recall (Section 1.11) that if a joint bivariate pdf is *normal*, so are its marginals and conditionals. This property extends to multivariate normal pdf's, so there is a class of *Gaussian random processes* for which *all* the pdf's are normal. In such a case, we can write down any particular pdf if we know its means and covariances.

The other simple class of random processes occurs when $X(t_1)$ is independent of $X(t_2)$ for all times t_1, t_2 (for example, the Bernoulli process). Then all the joint pdf's are simply products of the first-order pdf's. (Interpret Equation 2.2 for independent $X(t_1)$, $X(t_2)$.)

Of course, we can't expect to have knowledge of all of these pdf's. In practice, we are typically interested in predicting the value of a random process at some time, perhaps using some information we have gleaned from measurements at other times. As we shall see, the construction of an important class of *optimal linear estimators* relies only on knowledge of the means and correlations of the process, as specified in the following tabulation:

SUMMARY: THE FIRST AND SECOND
MOMENTS OF RANDOM PROCESSES

$$\text{mean of } X(t) = E\{X(t)\} = \int_{-\infty}^{\infty} x f_{X(t)}(x) dx = \text{``}\mu_X(t)\text{''}$$

autocorrelation of $X(t)$ = correlation between $X(t_1)$ and $X(t_2)$

$$= E\{X(t_1)X(t_2)\}$$

$$= \int_{-\infty}^{\infty} \int_{-\infty}^{\infty} x_1 x_2 f_{X(t_1), X(t_2)}(x_1, x_2) dx_1 dx_2$$

$$= \text{``}R_X(t_1, t_2)\text{''}$$

autocovariance of $X(t)$ = correlation between $[X(t_1) - \mu_X(t_1)]$ and $[X(t_2) - \mu_X(t_2)]$

$$= E\{[X(t_1) - \mu_X(t_1)][X(t_2) - \mu_X(t_2)]\}$$

$$= \int_{-\infty}^{\infty} \int_{-\infty}^{\infty} [x_1 - \mu_X(t_1)][x_2 - \mu_X(t_2)] f_{X(t_1), X(t_2)}(x_1, x_2) dx_1 dx_2$$

$$= \text{``}C_X(t_1, t_2)\text{''}$$

crosscorrelation of two random processes $X(t)$ and $Y(t)$

$$= \text{correlation between } X(t_1) \text{ and } Y(t_2)$$

$$= E\{X(t_1)Y(t_2)\}$$

$$= \int_{-\infty}^{\infty} \int_{-\infty}^{\infty} x_1 y_2 f_{X(t_1), Y(t_2)}(x_1, y_2) dx_1 dy_2$$

$$= \text{``}R_{XY}(t_1, t_2)\text{''}$$

crosscovariance of two random processes $X(t)$ and $Y(t)$

$$= \text{correlation between } [X(t_1) - \mu_X(t_1)] \text{ and } [Y(t_2) - \mu_Y(t_2)]$$

$$= E\{[X(t_1) - \mu_X(t_1)][Y(t_2) - \mu_Y(t_2)]\}$$

$$= \int_{-\infty}^{\infty} \int_{-\infty}^{\infty} [x_1 - \mu_X(t_1)][y_2 - \mu_Y(t_2)] f_{X(t_1), Y(t_2)}(x_1, y_2) dx_1 dy_2$$

$$= \text{``}C_{XY}(t_1, t_2)\text{''}$$

crosscorrelation coefficient

$$= \frac{C_{XY}(t_1, t_2)}{\sqrt{C_X(t_1, t_1) C_Y(t_2, t_2)}} = \frac{C_{XY}(t_1, t_2)}{\sigma_X(t_1) \sigma_Y(t_2)}$$

$$= \text{``}\rho_{XY}(t_1, t_2)\text{''}$$

As always, the covariances equal the correlations for zero-mean processes. Otherwise, the familiar *second-moment identities* hold (Problem 12):

$$R_X(t_1, t_2) = C_X(t_1, t_2) + \mu_X(t_1)\,\mu_X(t_2)$$
$$R_{XY}(t_1, t_2) = C_{XY}(t_1, t_2) + \mu_X(t_1)\,\mu_Y(t_2). \tag{2.3}$$

Note that if $t_1 = t_2$, the autocovariance equals the variance of $X(t)$ at that time:

$$C_X(t, t) = \sigma_X^2(t). \tag{2.4}$$

Example 2.11

For the Bernoulli process (Figure 2.12), the probability of 0 (or 1) on a given toss is 0.5, and the probability of 0 (or 1) on one toss and 0 (or 1) on another toss is $0.5^2 = 0.25$. Thus, from the definitions we have,

$$\mu_X(t) = (0.5) \times 0 + (0.5) \times 1 = 0.5 \quad \text{if } t = 0, \pm 1, \pm 2, \ldots \quad (0 \text{ otherwise})$$

$$R_X(t_1, t_2) = \frac{1}{4}(0 \times 0) + \frac{1}{4}(0 \times 1) + \frac{1}{4}(1 \times 0) + \frac{1}{4}(1 \times 1) = \frac{1}{4} \quad \text{if } t_1 \neq t_2, \quad \text{both integers}$$

$$R_X(t_1, t_2) = \frac{1}{2}(0 \times 0) + \frac{1}{2}(1 \times 1) = \frac{1}{2} \quad \text{if } t_1 = t_2, \quad \text{both integers}$$

$$R_X(t_1, t_2) = 0 \quad \text{otherwise.}$$

$$C_X(t_1, t_2) = \frac{1}{4} - \left(\frac{1}{2}\right)^2 = 0 \quad \text{if } t_1 \neq t_2, \quad \text{both integers}$$

$$C_X(t_1, t_2) = \frac{1}{2} - \left(\frac{1}{2}\right)^2 = \frac{1}{4} \quad \text{if } t_1 = t_2, \quad \text{both integers}$$

$$C_X(t_1, t_2) = 0 \quad \text{otherwise.}$$

Did you catch the change in the autocorrelations when the times are equal?

Example 2.12

Suppose $X(t)$ is a zero-mean ($\mu_X(t) \equiv 0$) Gaussian process with autocorrelation function $R_X(t_1, t_2) = 4e^{-|t_1 - t_2|}$.

i. What is the probability that the value of $X(7)$ lies between 6 and 7?
ii. If $X(6)$ is measured and found to have the value 1, what is the probability that the value of $X(7)$ lies between 6 and 7?
iii. What is the probability that $X(6)$ and $X(7)$ both lie between 0 and 7?

Solution

These questions sound formidable, but they are much easier than they appear.

   i. $X(7)$ is normal, so its pdf is normal with mean and standard deviation

$$\mu_X(7) = 0, \sigma_X(7) = \sqrt{C_X(7,7)} = \sqrt{R_X(7,7) - \mu_X(7)^2} = \sqrt{4e^{-|7-7|}} = 2,$$

and the probability is the integral of $N(0,2)$ (Section 1.9) from 6 to 7. (Answer: 0.0011)

   ii. According to Section 1.12 the conditional mean and standard deviation for a normal bivariate distribution are related to the correlation coefficient by

$$\rho = \text{cov}\frac{(X,Y)}{(\sigma_X \sigma_y)}$$

$$\mu_{X|Y} = \mu_X + \rho\frac{\sigma_X}{\sigma_Y}(y - \mu_Y)$$

$$\sigma_{X|Y} = \sigma_X \sqrt{1 - \rho^2}.$$

Interpreting "X" as $X(7)$, "Y" as $X(6)$, and "covariance" as "autocorrelation" (for zero-mean processes), we find the following:

$$\mu_{X(7)} = \mu_{X(6)} = 0, \sigma_{X(7)} = \sigma_{X(6)} = \sqrt{4e^0} = 2.$$

$$\rho_{X(7)X(6)} = \frac{4e^{-|7-6|}}{2 \times 2} = e^{-1} \approx 0.3769.$$

$$\mu_{X(7)|X(6)} = 0 + e^{-1}\frac{2}{2}(1 - 0) = e^{-1} \approx 0.3769.$$

$$\sigma_{X(7)|X(6)} = 2\sqrt{1 - e^{-2}} \approx 1.8597.$$

The probability is the integral of $N(2, 1.8597)$ from 6 to 7. (Answer: 0.0010)

   iii. Putting these values into the bivariate normal pdf of Section 1.12 and integrating over $0 < x < 7$ and $0 < y < 7$, we obtain 0.3095.

Example 2.13

Describe the mean and autocorrelation of a switch that is turned on at a random time between $t = 0$ and $t = 1$. That is, the process $X(t)$ is zero for $t < c$ and one for $t > c$, where c is a random variable uniformly distributed in $[0,1]$ (Figure 2.14).

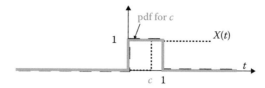

FIGURE 2.14 Random switching function.

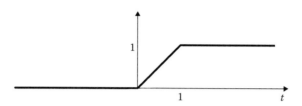

FIGURE 2.15 Probability that $X(t) = 1$ (*and*, the mean of $X(t)$).

Solution

$X(t)$ is certainly zero if $t < 0$ and certainly 1 if $t > 1$. If $0 < t < 1$, the probability that the switching time c precedes t is, in fact, t; and in such a case $X(t) = 1$. Thus the probability that $X(t) = 1$ is given by the graph in Figure 2.15; and since the mean of $X(t)$ equals 0 times probability that $X(t)$ is zero plus 1 times the probability that it is one, the same graph gives the mean.

To calculate the autocorrelation, note that $X(t_1)X(t_2)$ is zero unless both t_1 and t_2 exceed c. So if the minimum of (t_1, t_2) is less than zero, the probability of $X(t_1)X(t_2) = 1$ is 0; if the minimum exceeds one, the probability is 1, and otherwise the probability *is* min(t_1, t_2). Again, the expected value of $X(t_1)X(t_2)$, that is, the autocorrelation equals this probability.

We shall turn to the problem of determining the correlations for specific random processes in Chapters 3 and 4.

2.3 PREDICTION: THE STATISTICIAN'S TASK

Why does anyone study probability and statistics? To predict the weather, the stock market, the World Series outcome, the reliability of a circuit, and the lifetime of a machine, we seek a rational way to formulate a guess about something that is uncertain. To take a specific situation, let's say we want to predict the temperature in Boston, in degrees Celsius, on a given day.

If we have no information at all, any number between −30°C and 44°C would be a valid guess. With a little research, we learn that the average temperature in Boston is 11°C so *that* number would be a wiser prediction. We hardly need a statistician for that!

But let's suppose we *know* that the temperature in New York City is 37°C on the given day. That means it's likely to be a lot hotter than 11°C in Boston. Is there a scientific way to incorporate the NYC information to get a better prediction for Boston? That would be valuable, and such *rational predictions* are the essence of the science of statistics.

Let's mentally explore this problem, and see what kind of data we would need in order to exploit the information about New York's temperature. We propose to predict the Boston temperature, B, from a formula that involves the NYC temperature, N. Probably the simplest formula would be

$$\hat{B} = \alpha + \beta N, \tag{2.5}$$

where we use the symbol $\hat{B}$ to denote the *predictor* of B.

How do we choose the best values to use for the coefficients α and β, and what type of data do we need to evaluate them? The predictor is not going to be correct every time, of course, but can we make the error $B - \hat{B} = B - (\alpha + \beta N)$ small—"in the long run," say? Of course, we mean "small, in magnitude" here (we don't want to reward large negative errors), so let's see what would be required to minimize the mean *squared* error (MSE)*

$$MSE \equiv E\left\{\left(B - \hat{B}\right)^2\right\} = E\left\{\left[B - (\alpha + \beta N)\right]^2\right\}. \tag{2.6}$$

Expanding this expression yields

$$E\left\{\left[B - (\alpha + \beta N)\right]^2\right\} = E\left\{B^2 + \alpha^2 + \beta^2 N^2 - 2\alpha B - 2\beta BN + 2\alpha\beta N\right\}$$
$$= E\left\{B^2\right\} + \alpha^2 + \beta^2 E\left\{N^2\right\} - 2\alpha E\left\{B\right\} - 2\beta E\left\{BN\right\} + 2\alpha\beta E\left\{N\right\}. \tag{2.7}$$

The *least* **mean squared error** (LMSE) will occur where the partial derivatives with respect to the coefficients are zero:

$$\frac{\partial}{\partial \alpha} E\{[B - (\alpha + \beta N)]^2\} = 2\alpha - 2E\left\{B\right\} + 2\beta E\left\{N\right\} = 0$$

$$\frac{\partial}{\partial \beta} E\{[B - (\alpha + \beta N)]^2\} = 2\beta E\left\{N^2\right\} - 2E\left\{BN\right\} + 2\alpha E\left\{N\right\} = 0,$$

which we reorganize as the system

$$\alpha + E\left\{N\right\}\beta = E\left\{B\right\}, \quad E\left\{N\right\}\alpha + E\left\{N^2\right\}\beta = E\left\{BN\right\},$$

---

* Of course the obvious strategy would be to minimize the mean absolute value of the error, $E\left\{\left|B - \hat{B}\right|\right\} = E\left\{\left|B - (a + \beta N)\right|\right\}$, but a little calculation will convince you that this would be much harder than minimizing the MSE.

and solve to get

$$\alpha = \frac{E\{B\}E\{N^2\} - E\{N\}E\{BN\}}{E\{N^2\} - E\{N\}^2} = \mu_B - \rho\frac{\sigma_B}{\sigma_N}\mu_N, \beta = \frac{E\{BN\} - E\{N\}E\{B\}}{E\{N^2\} - E\{N\}^2} = \rho\frac{\sigma_B}{\sigma_N}$$

(2.8)

(using the relations in Section 1.11). So we see the construction of the least mean squared error predictor requires only the first and second moments (including the cross-correlation) of the variable being predicted and of the variable used in the predictor.

Our basic task is to learn how to predict the values of random processes. From what we have seen, we anticipate that this will involve evaluating their first and second moments and constructing the predictor. Chapters 3 and 4 address the former objective and Chapters 5 and 6 the latter.

In closing this chapter, we take a final glance at the LMSE prediction and ask how big *is* the LMSE. We would expect it to be small if the temperatures B and N are highly correlated, but if there is no correlation, the value of N would be of no help in predicting B and the prediction error would merely reflect the "spread" in the random variable B. So despite a lot of algebra (Problem 26), the formula for the LMSE is simple; it bears this out and holds little surprise:

$$E\left\{\left[B - (\alpha + \beta N)\right]^2\right\} = \left(1 - \rho^2\right)\sigma_B^2$$

$$= 0 \quad \text{if } \rho = \pm 1,$$ (2.9)

$$= \sigma_B^2 \quad \text{if } \rho = 0.$$

EXERCISES

Interpret all sequences $X(n)$ herein as starting from $X(0)$.

SECTION 2.2

1. A coin is flipped each second. When it is heads, the random variable $X(n)$ takes the value $+1$; when it is tails then $X(n) = -1$. What are the mean and autocorrelation of $X(n)$?

2. The random process $X(n)$ is created by spinning a dial that selects numbers $X(n) = [1, 2, \text{ or } 3]$ with equal probability. What are the mean, autocorrelation, and autocovariance for this process?

3. There are three signal generators in a box. The first one puts out the sequence {1 1 1 1 1 1 1 1 1 1 1 1 ...}. The second puts out {2 0 2 0 2 0 2 0 2 0 2 0 2 ...}. The third puts out {0 2 0 2 0 2 0 2 0 2 0 2 0 ...}. One of these generators is selected at random (equally likely). What are the mean $\mu(n)$, autocorrelation $R_X(m, n)$, and autocovariance $C_X(m, n)$ of the resulting sequence for $m = 3$ and $n = 4$?

4. The random process $X(n)$ is determined by a single flip of a fair coin. If the coin shows heads, then the process is a series of ones: {1 1 1 1 1 1 1 1 1 ...}. If the coin shows tails, then $X(n) = 1/(n+1)$: {1 1/2 1/3 1/4 1/5 ...}. What are the mean and the autocorrelation?

5. The random process $X(n)$ is created by starting with the *deterministic* sequence $Y(n) = (-1)^n$: $\{1 \ -1 \ 1 \ -1 \ 1 \ -1 \ 1 \ -1 \ ...\}$ and adding ± 1 to each term according to the outcome of a coin flip (+1 for heads, −1 for tails). So if the sequence of (independent) coin flips turned out to be HHTTHT..., the random process outcomes would be

	$(1 + 1)$	$(-1 + 1)$	$(1 - 1)$	$(-1 - 1)$	$(1 + 1)$	$(-1 - 1)$	...
or	2	0	0	−2	2	−2	...

What are the mean, autocorrelation, and autocovariance?

6. The random process $\{X(n)\}$ is created by starting with the *nonrandom* repeating sequence $\{1, 2, 3, 1, 2, 3, 1, 2, 3, ...\}$ and adding to each term the outcome of a throw of a die (each integer 1 through 6 is equally likely). So if the sequence of (independent) throws turned out to be $\{4 \ 2 \ 3 \ 4 \ 1 \ 5 \ ...\}$, the random process outcomes would be $\{5 \ 4 \ 6 \ 5 \ 3 \ 8 \ ....\}$. What are the following autocorrelations: $R_X(0, 0)$, $R_X(1, 1)$, $R_X(2, 2)$, $R_X(0, 1)$, $R_X(1, 0)$, $R_X(0, 2)$?

7. A sequence $X(n)$ of coin flips is recorded in the following manner. For trials #0, 2, 4, 6, 8, ... and all even-numbered trials, the outcome of the experiment is 0 if the coin reads heads, and 1 if the coin reads tails. On odd-numbered trials #1, 3, 5, 7, 9, ..., the outcome is 1 if the coin reads heads, and 0 if the coin reads tails. Consider the following as an example:

Trial	0	1	2	3	4	5	6	7	8	9
Coin	H	H	H	T	H	T	T	H	T	T
X	0	1	0	0	0	0	1	1	1	0

What are the mean, autocorrelation, and autocovariance of X?

8. Suppose the random process $X(t) = e^{at}$ is a family of exponentials depending on the value of a random variable a with pdf $f_a(a)$. Express the mean, the autocorrelation, and the pdf $f_{X(t)}(x)$ of $X(t)$ in terms of the pdf $f_a(a)$ of a.

9. A fair coin is flipped one time. Define the process $X(t)$ as follows: $X(t) = \cos \pi t$ if *heads* shows, and $X(t) = t$ if *tails* shows. Find $E\{X(t)\}$ for $t = 0.25$, $t = 0.5$, and $t = 1$.

10. Express the mean and autocorrelation of $Y(t) = X(t) + 5$ in terms of those of $X(t)$.

11. Let $p(t)$ be a periodic square wave as illustrated in Figure 2.16c. Suppose a random process is created according to $X(t) = p(t - \tau)$, where τ is a random variable uniformly distributed over $(0, T)$. Find $f_{X(t)}(x)$, $\mu_{X(t)}$, and $R_{X(t)}(t_1, t_2)$.

12. Prove the second-moment identities (2.3) in Section 2.2.

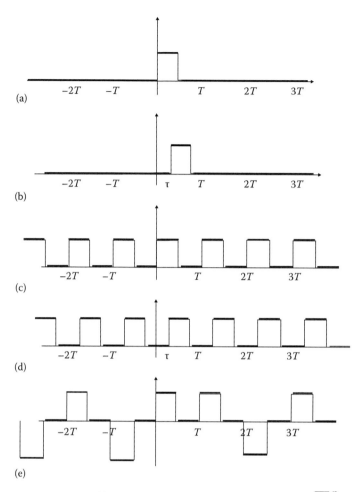

FIGURE 2.16 (a) Pulse $p(t)$. (b) Delayed pulse $p(t-\tau)$. (c) Pulse train $\sum_{n=-\infty}^{\infty} p(t-nT)$, (d) Delayed pulse train $\sum_{n=-\infty}^{\infty} p(t-nT-t)$. (e) Pulse code modulation for $A_n = \dots -1\ 1\ -1$ $1\ 1\ -1\ 1 \dots$.

13. Let $X(t)$ and $Y(t)$ be independent processes.
  a. Determine the autocorrelation functions of $Z_1(t) = X(t) + Y(t)$ and $Z_2(t) = X(t) - Y(t)$ in terms of the moments of X and Y.
  b. Determine the crosscorrelation function of $Z_1(t)$ and $Z_2(t)$.
  c. How do these formulas simplify if X and Y have identical means and autocorrelations?

14. Suppose the energy collected each day by a solar panel is normal $N(\mu, \sigma)$ and independent of the energies collected on other days and the collector is ideal (100% efficient, lossless). What are the mean and autocorrelation of the *accumulated* energy from day #1 to day #n?

15. Find the autocorrelation for the random process $X(t) = \delta(t-c)$ when c is uniformly distributed in the interval (0, 1).

16. Suppose $X(n)$ is a discrete zero-mean random process and for each n, $X(n)$ is independent of all other $X(j)$ except $X(n \pm 1)$. The autocorrelation is given by $R_X(n, n) = 2$, $R_X(n, n-1) = R_X(n, n+1) = 1$. What are the mean and autocorrelation of the "moving average" sequence $Y(n) = \dfrac{X(n) + X(n-1) + X(n-2)}{3}$?

17. If $R_X(t_1, t_2)$ is a function only of $t_1 - t_2$, what is the autocorrelation of $Y(t) \equiv X(t) - X(t - T)$ (for constant T)?

18. (**Pulse code modulation**. *Refer to Figure* 2.16.) Let A_n be a random sequence determined by coin flips, $+1$ for heads and -1 for tails, and let $p(t)$ be the rectangular pulse depicted in Figure 2.16a.

    $X(t)$ is the random process formed by modulating the pulse train generated by $p(t)$ by the sequence $\{A_n\}$: $X(t) = \sum_{n=-\infty}^{\infty} A_n\, p(t - nT)$. A possible outcome is depicted in Figure 2.16e.

    a. What are the mean and autocovariance of $X(t)$?
    b. What are the mean and autocovariance of a randomly delayed version of $X(t)$: $Y(t) = X(t - \tau)$, where τ is uniformly distributed over the interval $[0, T]$ and is statistically independent of the $\{A_n\}$?

19. Rework Problem 18 if $\{A_n\}$ are independent, but take values $+a$ with probability p and $+b$ with probability $(1 - p)$.

In communications engineering, it is common to combine message signals with sinusoids ($\cos\omega t, \sin\omega t$) to facilitate their transmission. The solutions to Problems 20–22 are facilitated by the trigonometric identities:

$$\cos A \cos B = \frac{\cos(A - B) + \cos(A + B)}{2}.$$

$$\sin A \sin B = \frac{\cos(A - B) - \cos(A + B)}{2}.$$

$$\sin A \cos B = \frac{\sin(A - B) + \sin(A + B)}{2}.$$

20. A random process is given by $X(t) = A\cos\omega t + B\sin\omega t$, where A and B are independent zero mean random variables.
    a. Find the mean $\mu_X(t)$.
    b. Find the autocorrelation function $R_X(t_1, t_2)$.
    c. Under what conditions on the variances of A and B does $R_X(t_1, t_2)$ depend only on $(t_1 - t_2)$?

21. (**Quadrature amplitude modulation**) Consider the signals $Z_1(t) = X(t)\cos\omega t$ and $Z_2(t) = X(t)\cos\omega t + Y(t)\sin\omega t$, where $X(t)$ and $Y(t)$ are zero-mean independent processes with identical autocorrelation functions $R_X = R_Y$. Determine $R_{Z_1}(t_1, t_2)$ and $R_{Z_2}(t_1, t_2)$ in terms of $R_X(t_1, t_2)$. Show that if $R_X(t_1, t_2)$ depends only on $(t_1 - t_2)$, so does $R_{Z_2}(t_1, t_2)$. (How about $R_{Z_1}(t_1, t_2)$?)

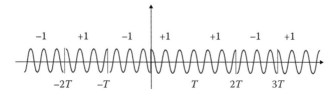

FIGURE 2.17   Binary phase shift keying for $A_n = \ldots$ -1 1 -1 1 1 -1 $1 \ldots$

22. (**Binary phase shift keying**) The ± 1 coin flip sequence $\{A_n\}$ described in Problem 18 is encoded for transmission as a $(\pm \pi/2)$ phase shift in a sine wave:

$$X(t) = \sin\left(\omega t + A_n \pi/2\right) = \begin{cases} \cos \omega t \ \text{if } A_n = +1 \\ -\cos \omega t \ \text{if } A_n = -1 \end{cases} \quad \text{for } nT \le t < (n+1)T$$

with $T \gg 2\pi/\omega$ (see Figure 2.17). What are the mean and autocorrelation of $X(t)$? Be careful to distinguish between the cases when t_1 and t_2 lie in different, or the same, intervals.

23. Assume $X(t)$ is a zero-mean Gaussian process with autocorrelation function $R_X\left(t_1, t_2\right) = e^{-|t_1 - t_2|/2}$.

   a.   What is the probability that the value of $X(7)$ lies between -1 and 2?

   b.   If $X(6)$ is measured and found to have the value 2, what is the probability that the value of $X(7)$ lies between -1 and 2?

   c.   What is the probability that $X(6)$ and $X(7)$ both lie between -1 and 2? (Consult Section 1.10 for MATLAB codes for numerical integration.)

24. Suppose that $X(t)$ is a zero-mean Gaussian process with autocorrelation function
$$R_X\left(t_1, t_2\right) = \frac{\sin \pi\left(t_1 - t_2\right)}{t_1 - t_2}.$$

   a.   What is the standard deviation of $X(1.5)$?

   b.   Write out the formula for the joint probability density $f_{X(t_1)X(t_2)}\left(x_1, x_2\right)$ of $X\left(t_1\right)$ and $X\left(t_2\right)$.

   c.   Write out the formula for the marginal probability density $f_{X(t_1)}\left(x\right)$ of $X\left(t_1\right)$.

   d.   If $X(1.5)$ is determined to be 3, what are the (conditional) mean and standard deviation of $X(2)$?

   e.   Write out the formula for the conditional probability density $f_{X(2)|X(1.5)}\left(x_2 \mid x_1\right)$ for $X\left(2\right)$, given that $X(1.5) = 3$.

   f. $X(3)$ and $X(4)$ are independent, but $X(1)$ and $X(1.5)$ are not. Explain.

25. If $X(t)$ is a Gaussian random process whose first-order pdf $f_{X(t)}\left(x\right)$ has constant mean = 0 and whose autocorrelation is given by $R_X\left(t_1, t_2\right) = e^{-|t_1 - t_2|}$, what is its third-order joint pdf $f_{X(t_1), X(t_2), X(t_3)}\left(x_1, x_2, x_3\right)$?

SECTION 2.3

26. Prove the formula (2.9), Section 2.3, for the LMSE.

3 Analysis of Raw Data

As we have seen in Section 2.3, the least-mean-squared-error prediction of the outcomes of a random process relies on the knowledge of the process's first and second moments. How do we deduce these parameters?

In this chapter, we are going to study how to efficiently estimate the moments of a random process from a record of the process—a "run of data" or *time series* $X(t_1), X(t_2), \ldots, X(t_N)$. (The *theoretical* evaluation of moments for specific random process models is covered in Chapter 4.) For most of this chapter, we assume the data have been uniformly sampled (and analog-captured on a tape, for example) and thus treat the random process—or at least, its measured samples—as discrete. For simplicity, we'll frequently break the rules and write $X(1), X(2), \ldots, X(N)$ instead of $X(\Delta t), X(2\Delta t), \ldots, X(N\Delta t)$; the relation between $X(n), X(n\Delta t)$, and $X(t)$ will be clear from the context. In the final section we'll list the continuous-time versions of our formulas.

Whenever we try to predict the future from the past, we are making a couple of assumptions that aren't necessarily true—namely, that the statistical parameters (like means and correlations) won't change with time and that the data run we are using is typical of all of the possible realizations of the process. The opening section of this chapter is dedicated to a deeper examination of these assumptions.

3.1 STATIONARITY AND ERGODICITY

When we have no idea of the mechanisms driving a random process, our only hope is to try to predict its future from its past. Thus, we postulate that the expected value of $X(N + 1)$, say, is well approximated by the average of the first N values recorded in our data set:

$$E\{X(N+1)\} \approx \frac{X(1) + X(2) + \cdots + X(N)}{N}.$$

We also propose that the same average would approximate *all* of the future means:*

$$\frac{X(1) + X(2) + \cdots + X(N)}{N} \approx E\{X(N+1)\} = E\{X(N+2)\} = \cdots \quad (3.1)$$

---

* Indeed, according to the "averaging" strategy analyzed in Section 1.13, if all the variables $X(n)$ are independent and identically distributed, the mean squared error in (3.1) can be expected to diminish like $\sigma_X / \left(\sqrt{N}\right)$. But we would hardly expect the values of a random process to be independent!

Now, there are two underlying premises hidden in this formula, and we want to be explicit about them. First of all, we are presuming that the process is stationary, in the sense that

$$E\{X(m)\} = E\{X(n)\} \quad \text{for all } m, n \quad (\text{or } E\{X(t_1)\} = E\{X(t_2)\} \quad \text{for all } t_1, t_2). \quad (3.2)$$

More precisely, a process satisfying (3.2) is said to be *stationary in the mean*.

Clearly, a process with long-term growth trends (inflation in the stock market) or seasonal variations (the Boston temperature) is not stationary, but often corrections can be made to restore stationarity—like expressing prices in 1940 dollars or averaging temperatures annually.

The other premise underlying (3.1) is known as **ergodicity**. Ergodicity is best understood by looking at instances when it is *not* present. Recall the dc power supply described in Section 2.1, whose voltage is set to a random number every night, evenly distributed (say) between −1 and +1 V. The expected value of the voltage at 9 a.m. tomorrow, taking into account all of the possibilities, is 0. (It is equally likely to have been set to +a volts as to −a volts.) But if we average over measured values of the power supply voltage taken today, we'll be averaging a constant—and this constant will almost certainly not be zero. The statistics of the random dc power supply cannot be fathomed by studying the time series of a particular instance—which means that it is *not* ergodic.

We can say roughly that for an ergodic random process, all of the statistical possibilities for the variable at any time are reflected in the long-time history of a particular instance of the process. If $f_{X(t)}(x)$ is the pdf for the possible values of X at time t, then $E\{X(t)\} = \int_{-\infty}^{\infty} x f_{X(t)}(x) \, dx$; and if we can reasonably expect this to be approximately $(X(1) + X(2) + \cdots + X(N))/N$ for a sufficiently long observed record of the process, then the process is ergodic (specifically, ergodic in the mean). Clearly we are struggling to make a statement about limits here, but we shelve the rigorous discussion until Section 3.2.

Continuing with our line of thought, to get an approximation of the second moments (i.e., the autocorrelation) of the process—say, $E\{X(1002)X(1001)\} \equiv R_X(1002, 1001)$—we take the average, over the recorded history, of the products of consecutive values of the process:

$$R_X(1002, 1001) \approx \frac{X(2)X(1) + X(3)X(2) + \cdots + X(1000)X(999)}{999}. \quad (3.3)$$

Clearly we are assuming stationarity and ergodicity again: specifically, *stationarity in autocorrelation* presumes that $E\{X(2)X(1)\} = E\{X(3)X(2)\} = \cdots = E\{X(1000)X(999)\}$. In fact, for stationary processes the autocorrelation* is a function of the time *difference* only[†]:

---

* The autocorrelation of a stationary process has some interesting mathematical properties; see Problem 3.

[†] Like most authors, we don't bother to introduce a special symbol for the autocorrelation of a stationary process; we simply write R_X (and C_X for autocovariance) in (3.4) with only *one* argument—namely, the *difference* of the two times that would normally be listed. Note that the magnitude $|n_1 - n_2|$ is appropriate since there is no difference between $E\{X(n_1)X(n_2)\}$ and $E\{X(n_2)X(n_1)\}$.

$$R_X(n_1, n_2) = R_X(0, n_1 - n_2) = R_X(|n_1 - n_2|) \quad \left[\text{or } R_X(t_1, t_2) = R_X(|t_1 - t_2|) \right]. \quad (3.4)$$

Note that the stationarity properties (3.2) and (3.4) would be guaranteed if the pdf $f_{X(t)}(x)$ is independent of t, and all of the joint distributions $f_{X(t_1), X(t_2)}(x_1, x_2), f_{X(t_1), X(t_2), X(t_3)}(x_1, x_2, x_3), \ldots$ are invariant to an overall time shift, depending only on the time differences.* Advanced textbooks call such processes "strictly stationary." We reserve the nomenclature **wide-sense stationary** for less restricted processes with time-independent first moments (3.2), and second moments depending only on time differences (3.4).

The same formula (3.3) could be used to estimate $R_X(1003,1002)$, $R_X(1004,1003)$, etc. Other values of the autocorrelation are approximated by

$$R_X(1001,1001) = R_X(0) \approx \frac{X(1)X(1) + X(2)X(2) + \cdots + X(1000)X(1000)}{1000}$$

$$R_X(1003,1001) = R_X(2) \approx \frac{X(3)X(1) + X(4)X(2) + \cdots + X(1000)X(998)}{998}$$

$$R_X(1004,1001) = R_X(3) \approx \frac{X(4)X(1) + X(5)X(2) + \cdots + X(1000)X(997)}{997}, \quad (3.5)$$

and so on.

3.2 THE LIMIT CONCEPT IN RANDOM PROCESSES

In stating that statistical moments can be estimated from time averages (for ergodic processes), we have been purposely vague about the accuracy of the estimate. Clearly we expect averages over longer time periods to be more reliable than short-time averages, and the notion of a "limit" should come into play. However, limits in random processes do not enjoy the clear-cut certitude that we are accustomed to in calculus. For instance, the mean of $X(t)$ for the Bernoulli coin flip process[†] in Example 2.1 is clearly 1/2. And we would expect that if we ran an experiment consisting of, say, 1,000 coin flips, the sample average would be close to 0.5, while if we ran 1,000,000 trials, the average would be closer.

But there is no certainty in this expectation. It is *not impossible* than we just might get 1,000,000 heads, so that the sample average (one) is nowhere close to the "ensemble" mean (one-half). If we seek to mimic the "ε, δ, n" vernacular of calculus, we have to temper the assertions.

For example, in calculus we say that if a sequence $\{y_n\}$ approaches y as n tends to infinity, then we can ensure that $|y - y_n|$ will be less than any given $\varepsilon > 0$ by choosing

---

* Thus, $f_{X(t_1), X(t_2), X(t_3)}(x_1, x_2, x_3) = f_{X(t_1 - \tau), X(t_2 - \tau), X(t_3 - \tau)}(x_1, x_2, x_3)$ for all τ.

[†] Heads, 1; tails, 0.

n sufficiently large. But as the Bernoulli process shows, maybe we can *never* be *sure*, for a random variable. So we have to consider how to interpret statements like "$Y_n(t) \rightarrow Y(t)$":

i. Perhaps the *probability*, that $|Y_n(t) - Y(t)|$ exceeds any ε, approaches 0 as $n \rightarrow \infty$. This is called "convergence in probability."

ii. Perhaps each outcome $Y_n(t)$ eventually lies within any ε of $Y(t)$ (except possibly for a set of outcomes with probability zero). This is "convergence almost everywhere." or "convergence with probability one."

iii. Possibly the pdf's $f_{Y_n(t)}(y)$ of the $Y_n(t)$'s approach $f_{Y(t)}(y)$. This is "convergence in distribution."

iv. Maybe $Y_n(t)$ approaches $Y(t)$ "in the mean square," that is, $E\{|Y_n(t) - Y(t)|^2\}$ goes to 0. This is "mean-square convergence."

Each of these criteria is discussed in advanced texts. In fact a well-developed "mean-square calculus" has been developed with mean-square continuity, mean-square differentiability, etc. Here are some typical theorems that are not hard to prove:*

a. If the autocovariance (Section 2.2) $C_X(\tau) \equiv E\{[X(t) - \mu_X][X(t-\tau) - \mu_X]\}$

for a stationary random process $X(t)$ satisfies $\lim_{T \rightarrow \infty} \dfrac{1}{T} \int_{-T}^{T} \left(1 - \dfrac{|\tau|}{T}\right) C_X(\tau)\, d\tau = 0$,

then $E\{X(t)\} \equiv \mu_X$ equals the mean-square limit, as $T \rightarrow \infty$, of $\dfrac{1}{T} \int_{t-T}^{t} X(\tau)\, d\tau$.

(For stationary discrete processes, if $\lim_{N \rightarrow \infty} \dfrac{1}{N} \sum_{n=-N}^{N} \left(1 - \dfrac{|n|}{N}\right) C_X(n) = 0$,

then μ_X equals the mean-square limit, as $N \rightarrow \infty$, of $\dfrac{1}{N} \sum_{N \text{ values of } n} X(n)$.)

This theorem, of course, addresses ergodicity and justifies using longtime averages like (3.1).

For the autocorrelation, we crank the theorem up a notch.

b. If $X(t)$ is stationary and $Z_\tau(t)$ is the process $Z_\tau(t) \equiv X(x)X(t-\tau)$ with autocovariance $C_{Z_\tau}(v) = E\{(Z_\tau(t) - \mu_{Z_\tau})(Z_\tau(t-v) - \mu_{Z_\tau})\}$, and if

$$\lim_{T \rightarrow \infty} \frac{1}{T} \int_{-T}^{T} \left(1 - \frac{|v|}{T}\right) C_{Z_\tau}(v)\, dv = 0,$$

then $R_X(\tau) \equiv E\{X(t)X(t-\tau)\}$ equals the mean-square limit of $\dfrac{1}{T} \int_{t-T}^{t} X(t)X(t-\tau)\, d\tau$.

---

* Helstrom, C.W. 1991. *Probability and Stochastic Processes for Engineers*. New York: Macmillan.

(For stationary discrete processes, if $\lim\limits_{N\to\infty}\dfrac{1}{N}\sum\limits_{n=-N}^{N}\left(1-\dfrac{|n|}{N}\right)C_{Z_m}(n)=0$,

then $R_X(m)$ equals the mean square limit, as $N\to\infty$, of

$\dfrac{1}{N}\sum\limits_{N\text{ values of }n}X(n)\,X(n-m)$.)

So a legitimate vindication for using longtime averages like (3.5), and invoking ergodicity in autocorrelation, is expressed in this theorem. However, the labyrinthine autocovariance $C_{Z_\tau}(v)$ involves fourth moments of X, and the hypotheses in theorems (a) and (b) are seldom verified in practice.

Pragmatically, we usually *presume* ergodicity. Again, if we can't predict the future from the past, and we don't have a model, what else do we have to go on?

c. A stationary random process $X(t)$ has a mean-square derivative at time t if and only if $\partial^2 R_X(t_1,t_2)/\partial t_1\partial t_2$ exists at $t_1=t_2=t$.*

3.3   SPECTRAL METHODS FOR OBTAINING AUTOCORRELATIONS

Means of stationary processes are easy to estimate from data records; Equation 3.1 of Section 3.1 can be implemented on a computer in an eyeblink. Autocorrelations are another matter. A stationary process has one mean $E\{X\}$, but an infinite number of autocorrelations $R_X(n)$. In the next few sections, we will discuss a sophisticated, extensively studied methodology for estimating autocorrelations of ergodic random processes using a finite record of data drawn from one specific realization of the process.

To whet your appetite for the rather lengthy derivation, let me reveal the ending. The long calculations implied in (3.5) of Section 3.1 are finessed by simply taking the fast Fourier transform (FFT) of the data $X(n)$, squaring the magnitude, dividing by the number of data points, and taking the inverse FFT! Not only is this quick but, as you might expect, the appearance of Monsieur Fourier heralds some frequency-domain insight into the workings of stationary random processes.

Common implementations of the FFT presume that the number of data points is a power of 2. For convenience, we express the number of points as $2N$, and renumber them as $\{X(-N), X(-N+1), \ldots, X(-1), X(0), X(1), X(2), \ldots, X(N-1)\}$—regarding them (as Fourier would) as samples of a random process running from minus infinity to plus infinity.

With the assumption that the process X is stationary and ergodic, we have seen that its autocorrelation can be estimated by averaging products of its observed values in our data set:

$$\hat{R}_X(m)=\frac{1}{\text{Number of terms}}\sum_{n}X(n)X(n-m). \tag{3.6}$$

---

* We shall identify shot noise in Section 4.9 as the derivative of the Poisson random process, but there we will circumvent this theorem by employing a formal and nonrigorous discussion.

Before we examine this closely, note that the right hand member is very similar to the convolution operation $X \circ Y(m) \equiv \sum_{n=-\infty}^{\infty} X(n)Y(m-n)$ that arises in Fourier analysis.* Strictly speaking, Fourier theory only applies to sequences which are suitably "summable," and we will see that stationary random processes do not meet the criterion. However, we ignore this for now, writing some equations that are literally incorrect, just to motivate the mechanics of the formalism. The necessary adjustments to our calculations will be addressed in Section 3.5.

The *discrete time Fourier transform* (DTFT) of a sequence $\{X(n)\}$ is defined to be

$$\text{"DTFT"} \quad \tilde{X}(f) = \sum_{n=-\infty}^{\infty} X(n)e^{-j2\pi nf}, \quad -\frac{1}{2} < f < \frac{1}{2}, \tag{3.7}$$

and it comes with an inversion formula

$$X(n) = \int_{-1/2}^{1/2} \tilde{X}(f)e^{j2\pi nf}\,df. \tag{3.8}$$

The cosines in these superpositions are depicted in Figure 3.1. In the inversion formula (3.8), the original discrete sequence $X(n)$ is expressed as an integral—a "sum over a continuum"—of sinusoids in the t-domain, $\{\cos 2\pi ft + j\sin 2\pi ft\}$, embracing all frequencies f between $-1/2$ and $1/2$, and sampled at the integer points $t = n$. The cosines occurring in this "sum" are depicted in the left of the figure. In formula (3.7), the transform $\tilde{X}(f)$ is expressed as a discrete sum of sinusoids in the f-domain, $\{\cos 2\pi nf - j\sin 2\pi nf\}$, embracing (only) integer frequencies n from $-\infty$ to ∞. The cosines in this sum are depicted on the right.

The *Fourier convolution theorem* states that the product of the DTFTs of two sequences equals the DTFT of their convolution.

If $\tilde{X}(f) = \sum_{n=-\infty}^{\infty} X(n)e^{-j2\pi nf}$ and $\tilde{Y}(f) = \sum_{n=-\infty}^{\infty} Y(n)e^{-j2\pi nf}$ then

$$\tilde{X}(f)\tilde{Y}(f) = \sum_{m=-\infty}^{\infty} Z(m)e^{-i2\pi nf}, \tag{3.9}$$

where

$$Z(m) = \sum_{n=-\infty}^{\infty} X(n)Y(m-n). \tag{3.10}$$

In short, $\widetilde{XY} = \widetilde{X \circ Y}$.

---

* Books on Fourier theory abound. Your author's favorite is Kammler, D.W. 2008. *A First Course in Fourier Analysis.* Cambridge, U.K.: Cambridge University Press.

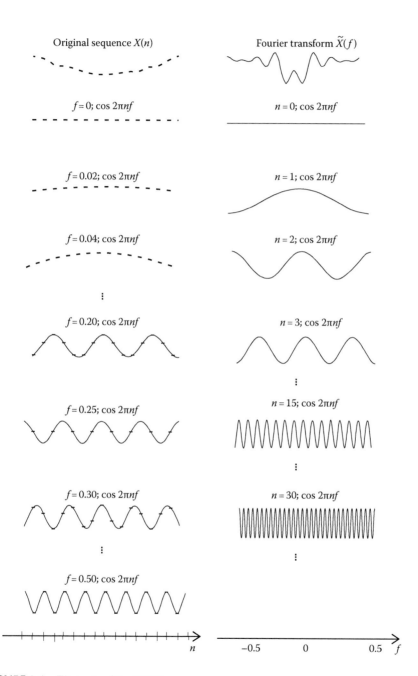

FIGURE 3.1 Elements of the DTFT.

Note that the summation index in (3.10) runs in opposite directions, whereas in (3.6) it runs in tandem. So we judiciously manipulate ((3.9) and (3.10)) by letting $Y(n)$ be the *time reversed* version of $X(n)$. As the following calculation demonstrates, the DTFT of $X(-n)$ is simply the complex conjugate of the DTFT of $X(n)$:

$$\tilde{Y}(f) = \sum_{n=-\infty}^{\infty} Y(n) e^{-j2\pi nf} = \sum_{n=-\infty}^{\infty} X(-n) e^{-j2\pi nf} = \sum_{n=-\infty}^{\infty} X(n) e^{+j2\pi nf} = \overline{\tilde{X}(f)}. \quad (3.11)$$

The convolution theorem thus tells us that

$$\tilde{X}(f)\tilde{Y}(f) = \tilde{X}(f)\overline{\tilde{X}(f)} \equiv \left|\tilde{X}^2(f)\right| = \sum_{m=-\infty}^{\infty} Z(m) e^{-j2\pi mf}$$

with $\quad Z(m) = \sum_{n=-\infty}^{\infty} X(n) Y(m-n) = \sum_{n=-\infty}^{\infty} X(n) X(n-m), \quad$ which has the form of our desired estimate (3.6)! In short, we wanted to calculate $(1/\text{number of terms})\left(\sum_{n} X(n) X(n-m)\right)$ and we can get something that looks very similar, $\sum_{n=-\infty}^{\infty} X(n) X(n-m)$, by taking the DTFT of $\{X(n)\}$, squaring its magnitude, and taking the inverse DTFT. This is certainly promising, since we can use the FFT algorithm for rapid evaluation of the DTFT.

3.4   INTERPRETATION OF THE DISCRETE TIME FOURIER TRANSFORM

The function $e^{j2\pi nf}$, as a function of the sample index n, is (of course) a complex sinusoid oscillating with frequency f cycles per sampling interval. (Only frequencies below half a cycle per sample need be considered; because of aliasing, frequencies higher than ½ are equivalent to lower frequencies when sampled. See Figure 3.2. Thus, we interpret the inverse Fourier transform $X(n) = \int_{-1/2}^{1/2} \tilde{X}(f) e^{j2\pi nf} df$ as expressing the process $X(n)$ as samples (at $t = n$) of a weighted linear combination of sinusoids $e^{j2\pi tf}$, summed over a continuum of frequencies.

To visualize this, in Figure 3.3, we display 20 samples of a sequence $X(n)$ that was synthesized as a (discrete) sum of (only) three sinusoids with frequencies 0.1, 0.3, and 0.4:

$$X(n) = \cos\left(2\pi n[0.1]\right) + \sin\left(2\pi n[0.3]\right) + \cos\left(2\pi n[0.4]\right)$$

$$= \frac{e^{j2\pi n[0.1]} + e^{-j2\pi n[0.1]}}{2} + \frac{e^{j2\pi n[0.3]} - e^{-j2\pi n[0.3]}}{2j} + \frac{e^{j2\pi n[0.4]} + e^{-j2\pi n[0.4]}}{2}$$

$$= \int_{-1/2}^{1/2} \left\{ \begin{array}{l} \dfrac{1}{2}\delta(f-0.1) + \dfrac{1}{2}\delta(f+0.1) + \dfrac{1}{2j}\delta(f-0.3) - \dfrac{1}{2j}\delta(f+0.3) \\[2mm] + \dfrac{1}{2}\delta(f-0.4) + \dfrac{1}{2}\delta(f+0.4) \end{array} \right\} e^{j2\pi nf} df.$$

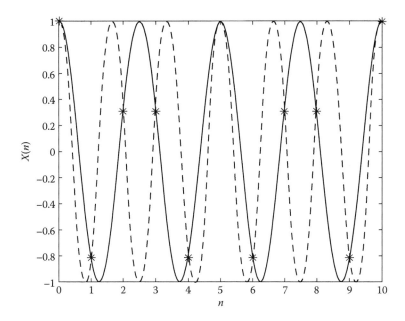

FIGURE 3.2   Aliasing. Solid curve, $X(n)=\cos 2\pi t(0.4)$; dashed curve, $X(n)=\cos 2\pi t(0.6)$.

The figure illustrates how we can get $X(n)$ by sampling a sum of sinusoids. Of course, in our situation, we *start* with the samples $X(n)$ and apply the DTFT to get the "recipe" for the sinusoids $\tilde{X}(f)$.

3.5   THE POWER SPECTRAL DENSITY

Although we observed in Section 3.3 that the autocorrelation estimate we desire

$$R_X(m) \approx \frac{1}{\text{Number of terms}} \sum_n X(n) X(n-m) \qquad (3.12)$$

looks very much like the output of the DTFT/magnitude-square/inverse-DTFT procedure

$$\sum_{n=-\infty}^{\infty} X(n) X(n-m) \qquad (3.13)$$

(the convolution of $X(n)$ and $X(-n)$), the difference is quite profound for a stationary random process. The basic problem is that for sums like (3.13) to converge, at the very least, the terms $X(n)$ should go to zero as $n \to \infty$. But if the process is stationary, the expected value of each $X(n)$ is the same—as is the expected value of $X(n)^2$. So virtually none of Fourier's equations are intelligible for stationary random processes.

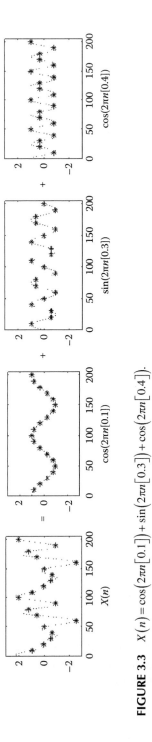

FIGURE 3.3 $X(n) = \cos\left(2\pi n[0.1]\right) + \sin\left(2\pi n[0.3]\right) + \cos\left(2\pi n[0.4]\right)$.

To underscore this, recall Parseval's theorem, which states that the DTFT of a (real) sequence $X(n)$ satisfies

$$\sum_{n=-\infty}^{\infty} X(n)^2 = \int_{-1/2}^{1/2} \left|\tilde{X}(f)\right|^2 df. \tag{3.14}$$

Now the sum in (3.14) has come to be known as the "energy" in the sequence by analogy with the other "square-law" formulas from physics:

i. The kinetic energy of a fluid equals $\iiint \frac{1}{2}\rho v^2 dx\, dy\, dz$.

ii. The potential energy of a mass-spring system equals $\frac{1}{2}kx^2$.

iii. The energy in an electromagnetic field equals $\frac{1}{2}\iiint \left\{\varepsilon|\mathbf{E}|^2 + \mu|\mathbf{H}|^2\right\} dx\, dy\, dz$.

Thus, we interpret the Parseval formula (3.14) as stating that the energy of a sequence is distributed over the frequency range $(-1/2, 1/2)$ with an *energy spectral density* equal to

$$\left|\tilde{X}(f)\right|^2.$$

However, this analogy is flawed in the case of stationary random processes. The *expected value* of their energy is infinite because $E\{X(n)^2\} = R_X(0)$, so that

$$E\left\{\sum_{n=-\infty}^{\infty} X(n)^2\right\} = \sum_{n=-\infty}^{\infty} R_X(0) = \infty \tag{3.15}$$

(unless $R_X(0) = 0$, corresponding to the trivial process consisting of all zeros). Stationary random processes have infinite energy (!), so they are not "summable" and the convergence of classical Fourier quantities like (3.13) is, as a statistician would say, unlikely.

A logical way out of this dilemma was formulated by Norbert Wiener, Alexander Khintchine, and their contemporaries.* They worked out a way to retool the Fourier machinery to accommodate infinite-energy sequences. To be sure, Wiener's

---

* Wiener, N. 1930. Generalized harmonic analysis. *Acta Mathematica* 55: 117–258; Khintchine, A. 1934. Korrelationstheorie der stationären stochastischen Prozesse. *Mathematische Annalen* 109: 604–615. Albert Einstein anticipated many of the issues in a largely overlooked paper in 1914: Einstein, A. 1914. Method for the determination of the statistical values of observations concerning quantities subject to irregular fluctuations. *Arch. Sci. Phys. et Natur.* 37(ser. 4): 254–256. See Yaglom, A.M. 1987. Einstein's 1914 paper on the Theory of Irregularly Fluctuating Series of Observations. *IEEE ASSP MAGAZINE* October: 7–11 and Jerison, D., Singer, I.M., and Stroock, D W. 1997. The legacy of Norbert Wiener: A centennial symposium. *Proceedings of Symposia in Pure Mathematics* 95. American Mathematical Society.

generalized harmonic analysis addresses processes $X(t)$ defined on the continuum $(-\infty < t < \infty)$, and it is quite profound and meticulous. But by focusing only on discrete processes $X(n)$, we will be able to concoct a (highly simplified) account of its reckoning.

If we truncate the troublesome (infinite) energy sum $\lim_{N \to \infty} E\left\{\sum_{n=-N}^{N-1} X(n)^2\right\}$ and divide by the number of terms before taking the limit, we obtain a convergent expression:

$$\lim_{N \to \infty} E\left\{\frac{1}{2N}\sum_{n=-N}^{N-1} X(n)^2\right\} = \lim_{N \to \infty} \frac{1}{2N}\sum_{n=-N}^{N-1} R_X(0) = \frac{2N}{2N} R_X(0) = R_X(0). \quad (3.16)$$

Since $2N$ sampling intervals (of length Δt) measure the *time* required to generate the samples $X(-N)$, $X(-N+1)$, ..., $X(N-1)$, the quantity in (3.16) has the units of energy divided by time: that is, *power*.* Thus, we say that *stationary random processes are finite-power processes*, and we express the power in the process by any of the equivalent formulas

$$\text{Power} = \lim_{N \to \infty} E\left\{\frac{1}{2N}\sum_{n=-N}^{N-1} X(n)^2\right\}$$

$$= R_X(0) \quad (\text{autocorrelation at zero time lag})$$

$$= E\left\{X(n)^2\right\} = \mu_X^2 + \sigma_X^2$$

$$= \lim_{N \to \infty} \frac{1}{2N}\sum_{n=-N}^{N-1} X(n)^2, \quad (3.17)$$

where we invoke ergodicity and interpret the extrapolation in the final formula as "mean-square limit" (Section 3.2). (Recall μ_X is the mean $E\{X(n)\}$ and σ_X is the standard deviation $E\left\{(X(n) - \mu_X)^2\right\}^{1/2}$, both independent of n for stationary processes.)

Let us apply the Fourier methodology to our data, maintaining a cautious vigilance toward infinities. We begin with the *truncated* sequence $X_N(n)$:

$$X_N(n) = \{\ldots, 0, 0, 0, 0, X(-N), X(-N+1), \ldots, X(N-1), 0, 0, \ldots\}. \quad (3.18)$$

Following the procedure suggested in Equations 3.12 and 3.13, we take its DTFT and square the magnitude:

$$\tilde{X}_N(f) = \sum_{n=-\infty}^{\infty} X_N(n)e^{-j2\pi nf} = \sum_{n=-N}^{N-1} X(n)e^{-j2\pi nf}, \quad (3.19)$$

---

* It is traditional to ignore the nuisance factor Δt. Equivalently, one could say we are using the sampling interval Δt as the time *unit*, restating "such-and-such per second" as "such-and-such per sample."

$$\left|\tilde{X}_N(f)\right|^2 = \tilde{X}_N(f)\overline{\tilde{X}_N(f)} = \left\{\sum_{n=-N}^{N-1}X(n)e^{-j2\pi nf}\right\}\left\{\sum_{p=-N}^{N-1}X(p)e^{+j2\pi pf}\right\}.$$

Now, we have to organize this collection of $2N\times 2N$ products into groups sharing a common value of $m \equiv p - n$. (If you try this for $N = 2$, you'll see the pattern; consult Figure 3.4 for the general case.) The result is

$$\left|\tilde{X}_N(f)\right|^2 = \sum_{m=-(2N-1)}^{0}e^{-j2\pi mf}\sum_{n=-N}^{N-1+m}X(n)X(n-m)+\sum_{m=1}^{2N-1}e^{-j2\pi mf}\sum_{n=-N+m}^{N-1}X(n)X(n-m).$$

(3.20)

For each m, the inner sum combines $2N - |m|$ terms of the form $X(n)X(n-m)$; thus, bearing in mind the estimator (3.12), we surreptitiously introduce this factor

$$\left|\tilde{X}_N(f)\right|^2 = \sum_{m=-(2N-1)}^{0}e^{-j2\pi mf}\left(2N-|m|\right)\frac{\sum_{n=-N}^{N-1+m}X(n)X(n-m)}{2N-|m|}$$

$$+\sum_{m=1}^{2N-1}e^{-j2\pi mf}\left(2N-|m|\right)\frac{\sum_{n=-N+m}^{N-1}X(n)X(n-m)}{2N-|m|}$$

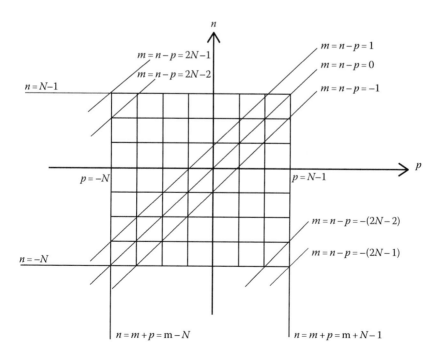

FIGURE 3.4   Change of indices in formula (3.20).

and heed the wisdom of (3.16) by dividing by the number of terms in the outer sum:

$$
\frac{\left|\tilde{X}_N(f)\right|^2}{2N} = \sum_{m=-(2N-1)}^{0} e^{-j2\pi mf} \frac{2N-|m|}{2N} \frac{\sum_{n=-N}^{N-1+m} X(n)X(n-m)}{2N-|m|}
$$

$$
+ \sum_{m=1}^{2N-1} e^{-j2\pi mf} \frac{2N-|m|}{2N} \frac{\sum_{n=-N+m}^{N-1} X(n)X(n-m)}{2N-|m|}
$$

$$
= \sum_{m=-(2N-1)}^{2N-1} c_{m,N} e^{-j2\pi mf}. \tag{3.21}
$$

If we fix m and let N approach infinity in the coefficient $c_{m,N}$ (violating mathematical protocol), we recognize a quotient $(2N-|m|)/2N$ that approaches one and a quotient that approaches $R_X(m)$ (in an ergodic sense). Making a leap of faith, we propose, then, that in some sense the limit of (3.21) is the DTFT of the autocorrelation, which we designate as "$S_X(f)$":

$$
\lim_{N\to\infty} \frac{\left|\tilde{X}_N(f)\right|^2}{2N} = \sum_{m=-\infty}^{\infty} R_X(m)e^{-j2\pi mf} = [\text{DTFT of } R_X] \equiv S_X(f).
$$

And in turn, the autocorrelation R_X is the inverse transform ((3.7) and (3.8)) of S_X. The significance of $S_X(f)$ is gleaned by noting that its inverse transform for $m=0$ is

$$
R_X(0) = \int_{-1/2}^{1/2} S_X(f)\,df. \tag{3.22}
$$

Therefore, the *power* in the random process (recall 3.17) can be regarded as distributed over the frequency interval $(-1/2, 1/2)$ with a **power spectral density** given by S_X:

$$
\text{Power spectral density } (PSD) = S_X(f) = \sum_{m=-\infty}^{\infty} R_X(m)e^{-j2\pi mf} = \lim_{N\to\infty} \frac{\left|\tilde{X}_N(f)\right|^2}{2N}. \tag{3.23}
$$

The rigorous validation of (3.22) and (3.23) resides in the work of Wiener and Khintchine; the equations express the "Wiener–Khintchine theory." Their insight replaced the unsuitable notion of *energy* with *power*, and thereby reincarnated the techniques of spectral analysis for finite-power sequences. Before we resume our pursuit of the Fourier-facilitated calculation of the autocorrelation, we digress in the next two sections to elaborate on the perspectives opened up by this extension of Fourier's work.*

3.6   INTERPRETATION OF THE POWER SPECTRAL DENSITY

We have identified a stationary random process as a finite-power sequence and showed that its power $E\{X(n)^2\} = R_X(0)$ is distributed in frequency according to the power spectral density S_X:

$$\text{Power} = R_X(0) = \int_{-1/2}^{1/2} S_X(f)\,df. \tag{3.24}$$

Recall that the classical Parseval identity for summable sequences states that their *energy* is distributed in frequency according to the energy spectral density:

$$\text{Energy} = \int_{-1/2}^{1/2} \left|\tilde{X}(f)\right|^2 df. \tag{3.25}$$

The *spectrum* of such a sequence is the set of frequencies for which $\tilde{X}(f)$ is not zero, and its *bandwidth* is the length of the largest interval that encompasses the spectrum. Stationary random processes do not have Fourier transforms, but the similarity of (3.25) and (3.24) suggests the analogous definitions:

The (power) **spectrum** of a stationary random process is the set of (positive) frequencies for which the power spectral density $S_X(f)$ is not zero, and its **bandwidth** is the length of the largest interval that encompasses its spectrum.

What does it mean for a zero-mean random process to have a broad spectrum? Or a narrow bandwidth?

  i. If a random process $X(n)$ has an extremely flat spectrum—say the PSD is identically one, $S_X(f) = 1$—then its autocorrelation is

$$R_X(n) = \int_{-1/2}^{1/2} 1 e^{j2\pi n f} df = \delta_{n0} = \begin{cases} 1 & \text{if } n = 0, \\ 0 & \text{otherwise}. \end{cases}$$

So different terms of the sequence $X(n)$ are completely *un*correlated; there is no way to predict one from another.

A random process that is unpredictable is regarded as "noise." In optics, light with a flat spectrum is white, so by analogy, such a random process is called "white noise." (If the power spectrum is broad but not completely flat,

---

* Although we have couched this discussion in the context of *ergodic* processes (for which it is validated by the Wiener–Khintchine theory), we can define $S_X(f)$ as the Fourier transform of the autocorrelation R_X for *any* stationary process and retain its interpretation as power spectral density.

the process may be called "colored noise.") A zero-mean Bernoulli process (Section 2.1) is white noise.

ii. If a random process has an extremely narrow bandwidth around zero frequency—as an extreme, say, the PSD is a delta function, $S_X(f) = \delta(f)$—then its autocorrelation, given by the inverse DTFT,

$$R_X(n) = \int_{-1/2}^{1/2} S_X(f) e^{j2\pi nf} df = \int_{-1/2}^{1/2} \delta(f) e^{j2\pi nf} df = 1,$$

is unity for every value of the delay n. If the process has mean zero, then, the correlation coefficient between every value of $X(n)$ is one:

$$\rho_{X(m),X(n)} = \frac{\mathrm{cov}\left(X(m), X(n)\right)}{\sigma_{X(m)}\sigma_{X(n)}} = \frac{R_X(m-n)}{\sqrt{R_X(0)}^2} = 1,$$

signifying that all members of the sequence are completely correlated. There is essentially no randomness in this narrowband process. The random DC power supply (Section 2.1) is completely correlated.

iii. If the power spectral density is concentrated around a single nonzero frequency—say, $S_X(f) = \delta(f - f_0) + \delta(f + f_0)$ (Equation 3.28 implies S_X must be an even function for a real process)—then $R_X(n) =$

$$\int_{-1/2}^{1/2} S_X(f) e^{j2\pi nf} df = \int_{-1/2}^{1/2} \left[\delta(f - f_0) + \delta(f + f_0)\right] e^{j2\pi nf} df = 2\cos 2\pi f_0 n$$

and

$$\rho_{X(m),X(n)} = \frac{\mathrm{cov}\left(X(m), X(n)\right)}{\sigma_{X(m)}\sigma_{X(n)}} = \frac{R_X(m-n)}{\sqrt{R_X(0)}^2} = \frac{2\cos 2\pi f_0(m-n)}{\sqrt{2(1)}^2}$$

$$= \cos 2\pi f_0(m-n).$$

If f_0 is, say, 1/8, then the term $X(n)$ is completely correlated ($\rho = 1$) with the terms $X(n \pm 8), X(n \pm 16), X(n \pm 24),\dots$. And the correlation among the data points between $X(0)$ and $X(8)$ varies like the cosine. The sine waves with random phase (Example 2.9) exhibit this power spectral density.

Let's elaborate on this. If a signal's Fourier transform is $\delta(f - f_0) + \delta(f + f_0)$, then the signal *is* a cosine of frequency f_0. It has maxima spaced $1/f_0$ apart, it has zeros spaced $1/2f_0$ apart, and it repeats with a period $1/f_0$. We could never make such assertions for a *random* signal. But if its PSD is $\delta(f - f_0) + \delta(f + f_0)$, then its autocorrelation—that is to say, its *averages* in the sense of $(1/\text{number of terms})\left(\sum_n X(n)X(n-m)\right)$—has the maxima, zeros, and periodicities of the cosine.

Now that we have a feeling for the relation between the power spectrum and the autocorrelation of a process, it is natural to ask an engineering question: To what extent can we reshape a random process's spectrum? The answer lies in digital filter theory and is discussed in Section 3.7.

3.7   ENGINEERING THE POWER SPECTRAL DENSITY

It is well known that when a digital filter is applied to a summable sequence, the DTFT of the output equals the DTFT of the input multiplied by the DTFT of the impulse response. It seems feasible, therefore, to try to employ a filter to shape the power spectral density of a stationary random process. The details are as follows.

A prototype linear time invariant system is depicted in Figure 3.5.

The **unit impulse response** of the system, $h(n)$, is defined to be the output that results when the input is a unit impulse applied at time $n = 0$, that is, δ_{n0} (equals one when $n = 0$, and zero otherwise). Then (due to time invariance) it is obvious that the response to a unit impulse applied m time units later, δ_{nm}, is the delayed $h(n-m)$; and by linearity the response to the input string $X(n)$ is given by the convolution

$$Y(n) = \sum_{m=-\infty}^{\infty} X(m) h(n-m). \tag{3.26}$$

By Fourier's convolution theorem (Equation 3.10, Section 3.3), the DTFTs of these sequences $Y(n)$, $h(n)$, and $X(n)$ are related by

$$\tilde{Y}(f) = \tilde{H}(f)\tilde{X}(f) \tag{3.27}$$

and $\tilde{H}(f)$ is known as the "transfer function" of the system.

Let's interpret this. When the input is represented as a superposition of sampled sinusoids

$$X(n) = \int_{-1/2}^{1/2} \tilde{X}(f) e^{j2\pi nf} df$$

(Equation 3.8 and Figure 3.1, Section 3.3), the intensity $\tilde{X}(f)$ of the sinusoid at frequency f is tempered by the factor $\tilde{H}(f)$ in Equation 3.27, so the output's representation is

$$Y(n) = \int_{-1/2}^{1/2} \tilde{Y}(f) e^{j2\pi nf} df = \int_{-1/2}^{1/2} \tilde{H}(f)\tilde{X}(f) e^{j2\pi nf} df.$$

The output has no frequency content at frequencies where $\tilde{H}(f) = 0$, its *stop band*. A depiction of a *narrow band-pass* filter is portrayed in Figure 3.6. $\tilde{H}(f)$ annihilates all frequencies, except those between 0.2 and 0.3 (positive and negative), *passing* the latter with little distortion.

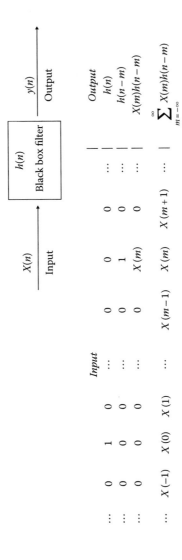

FIGURE 3.5 Linear time invariant system responses.

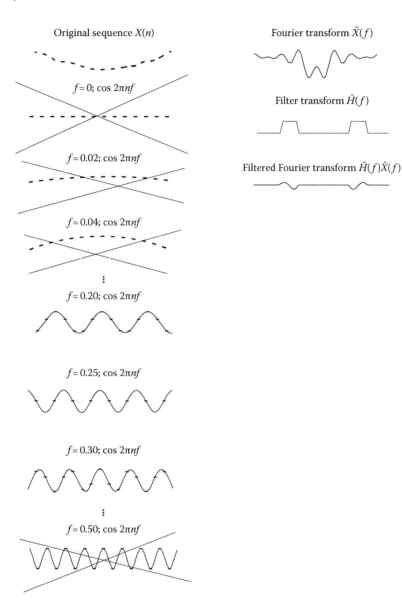

FIGURE 3.6   Narrow band-pass filter.

However, if $X(n)$ is a (*non*summable) stationary random process, then presumably $Y(n)$ is also. (This assumes the filter is sufficiently well behaved and any transient in the response has died out.)* Bearing in mind that X and Y are not summable and the Fourier theory is inapplicable, we ask what does the filter accomplish. We can resurrect the Fourier notions by shifting the focus to the autocorrelations.

---

* These issues are discussed in Section 4.2.

Observe that (3.26) implies $Y(0) = \sum_{p=-\infty}^{\infty} h(-p) X(p)$. Now multiply (3.26) by $Y(0)$ and take the expected values (to get autocorrelations). The expected value of

$$Y(n)Y(0) = \sum_{m=-\infty}^{\infty} h(n-m) X(m) Y(0) = \sum_{m=-\infty}^{\infty} h(n-m) X(m) \sum_{p=-\infty}^{\infty} h(-p) X(p)$$

is $R_Y(n) = \sum_{m=-\infty}^{\infty} h(n-m) \sum_{p=-\infty}^{\infty} h(-p) R_X(m-p)$; thus

$$R_Y(n) = \sum_{m=-\infty}^{\infty} h(n-m) Q(m) \quad \text{and} \quad Q(m) = \sum_{p=-\infty}^{\infty} h(-p) R_X(m-p). \quad (3.28)$$

Applying the convolution theorem to (3.28), we learn that the DTFT of R_Y—that is, the power spectral density S_Y—equals the DTFT of h times the DTFT of Q, and the latter equals the DTFT of R_X (namely S_X) times the DTFT of the "time-reversed" version of $h(p)$, which, as we noted in Equation 3.11 of Section 3.3, equals the complex conjugate of the DTFT of $h(p)$. In terms of power spectral densities, then, we have shown

$$S_Y(f) = \tilde{H}(f) S_X(f) \overline{\tilde{H}(f)} = \left| \tilde{H}(f) \right|^2 S_X(f). \quad (3.29)$$

Equation 3.29 spells out the possibilities for reshaping the power spectrum of a random process by digital filtering. Not only are the frequencies in a *summable* input signal annihilated where $\tilde{H}(f)$ is zero, they are also annihilated in the autocorrelation of a *random* stationary input signal; similarly, the frequencies where $\tilde{H}(f) = 1$ are undistorted both in summable signals and in the autocorrelations of random stationary signals. Figure 3.5 can be equally well regarded as depicting the action of a band-pass filter on a random signal, if we reinterpret $X(n)$ as the autocorrelation (and note that its Fourier transform—the power spectral density—can never be negative).

What are the possibilities? If S_X is narrow band, multiplying it by $\left| \tilde{H}(f) \right|^2$ will not change its shape very much. However, if the spectrum is *broadband*, like white noise, an appropriate $\tilde{H}(f)$ can create any S_Y; simply let $\left| \tilde{H}(f) \right|^2 = S_Y(f) / S_X(f)$. One could "whiten" a process by using a filter shaped like the reciprocal of the given S_X (unless the latter is zero). So, a good digital filter designer can tailor power spectral densities. In Section 4.4, we will study the autoregressive moving average procedure for indirectly designing systems with prescribed PSDs, by matching the corresponding autocorrelations.

We close by noting that the transfer function $\tilde{H}(f)$ of any linear time invariant system (Figure 3.4) can be easily evaluated, because the output $Y(n)$ equals $\tilde{H}(f) e^{j2\pi nf}$ when the input $X(n)$ is the sequence $X(n) = e^{j2\pi nf}$:

$$Y(n) = \sum_{m=-\infty}^{\infty} h(n-m)X(m) = \sum_{m=-\infty}^{\infty} h(n-m)e^{j2\pi mf}$$

$$= \sum_{m=-\infty}^{\infty} h(n-m)e^{-j2\pi(n-m)f}e^{j2\pi nf} = \hat{H}(f)e^{j2\pi nf}. \qquad (3.30)$$

Example 3.1

The input/output relation for a linear time-invariant system is described by the equation

$$Y(n) = 0.5Y(n-1) + 0.1Y(n-2) + 2X(n) + X(n-1). \qquad (3.31)$$

(In Section 4.3, this will be identified as an autoregressive-moving-average system of order (2,1).) If the input X is zero-mean white noise with power 5 units, what is the power spectral density of the output?

Solution

In the previous section, the power spectral density of white noise with *unit* power was identified as $S_{unit}(f) \equiv 1$, so our $X(n)$ has $S_X(f) \equiv 5$ (R_X is proportional to S_X; see Equation 3.23, Section 3.6). Substitution of $e^{j2\pi nf}$ into (3.31) yields

$$\tilde{H}(f)e^{j2\pi nf} = 0.5\tilde{H}(f)e^{j2\pi(n-1)f} + 0.1\tilde{H}(f)e^{j2\pi(n-2)f} + 2e^{j2\pi nf} + e^{j2\pi(n-1)f},$$

and solving for $\tilde{H}(f)$ is child's play.

$$\tilde{H}(f) = \frac{2 + e^{-j2\pi f}}{1 - 0.5e^{-j2\pi f} - 0.1e^{-j4\pi f}},$$

so (3.29) implies

$$S_Y(f) = \left| \frac{2 + e^{-j2\pi f}}{1 - 0.5e^{-j2\pi f} - 0.1e^{-j4\pi f}} \right|^2 5.$$

3.8   BACK TO ESTIMATING THE AUTOCORRELATION

The Wiener–Khintchine theory tells us that the autocorrelation and the power spectral density are Fourier transform pairs (Equations 3.22 and 3.23):

$$S_X(f) = \sum_{k=-\infty}^{\infty} R_X(k)e^{-j2\pi kf}, \quad R_X(k) = \int_{-1/2}^{1/2} S_X(f)e^{j2\pi kf} df. \qquad (3.32)$$

The form for S_X (Equation 3.21 and following, Section 3.5)

$$S_X(f) = \lim_{N \to \infty} P_N(f)(\text{mean square}), \quad P_N(f) = \frac{\left|\tilde{X}_N(f)\right|^2}{2N+1},$$

$$\tilde{X}_N(f) = \sum_{n=-N}^{N-1} X(n)e^{-j2\pi nf} \tag{3.33}$$

suggests that it can be estimated from sufficiently many data samples $X(-N)$, $X(-N + 1)$, ..., $X(N - 1)$. $P_N(f)$ in (3.33) is known as the "periodogram" of the data. Thus, the scheme for extracting the autocorrelation R_X is as follows:

i. Take the DTFT $\tilde{X}_N(f) = \sum_{n=-N}^{N-1} X(n)e^{-j2\pi nf}$ of a run of measurements $\{X(n), -N \leq n \leq N - 1\}$.

ii. Multiply this transform by its complex conjugate and divide by $2N$ (i.e., compute the periodogram).

iii. Take the inverse DTFT.

The result is an estimate of $R_X(k)$ for each k between 0 and $2N - 1$.

If the DTFT is implemented using the FFT, the number of distinct data points is usually given by a power of 2. The inverse DFTF, implemented via FFT, yields an equal number of estimates of $R_X(k)$.

After all this elegant theory, it is very disappointing to observe how poorly this autocorrelation estimate performs, even for large data sets.

For example, we simulated a particular $X(n)$ by synthesizing 1024 points for an ARMA random process, for which we *know* the autocorrelation $R_X(k)$. (Section 4.4 describes the methodology for constructing the signal.) The true autocorrelations are $R_X(0) = 4/3$, $R_X(1) = -2/3$, $R_X(2) = 1/3$, $R_X(3) = -1/6$, ..., and the true power spectral density is known to be $4/(5+4\cos 2\pi f)$. The simulated values of $X(n)$ appear random and stationary (Figure 3.7).

But despite the large value of N, the periodogram is very messy, compared to the theoretical PSD (Figure 3.8).

Specialists have concluded that the basic trouble here is that we have used all the data to compute only *one* estimate for each of the 1024 values of the periodogram. No averaging has been done to reduce the standard deviations of these 1024 estimates.

Much research in the last 50 years has gone into repairing this problem.* Here we shall only discuss one remedy, that of Bartlett.[†] Bartlett suggested we subdivide the data and make several different estimates of the power spectral density and average these estimates.

For the data at hand, we divided the $\{X(n)\}$ into 16 runs of 64 points each, yielding 16 periodograms. Of course, this only yields 64 points in the periodograms—that is,

---

* See Hayes, M.H. (1996). *Statistical Digital Signal Processing and Modeling*. New York: John Wiley & Sons.
[†] Bartlett, M.S. (1948). Smoothing periodograms from time-series with continuous spectra. *Nature* 161: 686–687.

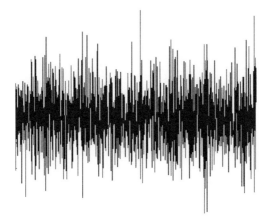

FIGURE 3.7   Data from ARMA (2,1) simulation.

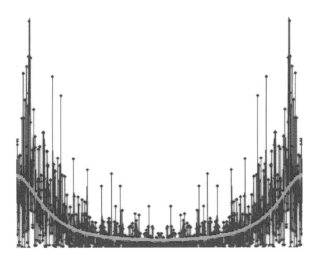

FIGURE 3.8   Periodogram with frequencies $-0.5 < f < 0.5$ (with true PSD).

estimates of the PSD for 64, rather than 1024, values of the frequency f. When the 16 periodograms are averaged, the resulting PSD estimate looks like Figure 3.9.

Figure 3.10 shows that the first five values of the autocorrelation calculated from the PSD estimate are reasonably close to the theoretical values.

The inaccuracies in the higher values of R_X are due to two "crimes":

i. We assumed that the quotient $(2N - |m|)/2N$ in (3.21) in Section 3.5 was close to its limit (one) as $N \to \infty$. But the index m runs between $\pm(2N-1)$, so this is certainly false for the higher $|m|$.

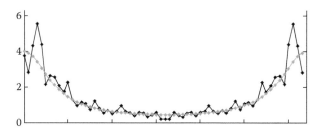

FIGURE 3.9 Bartlett PSD estimate, $-0.5 < f < 0.5$.

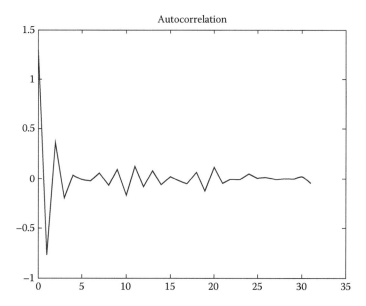

FIGURE 3.10 Autocorrelation estimates ($R_X(0) \equiv 4/3$, $R_X(1) \equiv -2/3$, $R_X(2) \equiv 1/3$, and $R_X(3) \equiv -1/6$).

ii. A finite data set, say $\{X(-N), X(-N + 1), \ldots, X(N - 1)\}$, provides $2N$ instances of $R_X(0)$—namely, $X(-N)^2$, $X(-N + 1)^2$, ..., $X(N - 1)^2$. But it only contains $(2N - 1)$ instances of $R_X(1)$—namely, $X(-N)X(-N + 1)$, $X(-N + 1)X(-N + 2)$, ..., $X(N - 2)X(N - 1)$; only $(2N - 2)$ instances of $R_X(2)$; and only one instance of $R_X(2N - 1)$. Thus, the estimates of $R_X(m)$ for the higher m values are less reliable.

So although the Wiener–Khintchine theory justifies identifying the limit-in-the-mean of the periodogram with the power spectral density (3.23), for any *finite* number of measurements $2N$, the estimates of the remote autocorrelations $R_X(m)$ are unreliable. Much of the advanced research in statistical spectral analysis has been dedicated to correcting for this.

ONLINE SOURCES

Fargues, M.P. Nonparametric spectrum estimation for stationary random signals. Naval Postgraduate School. Accessed June 24, 2016, http://faculty.nps.edu/fargues/teaching/ec4440/ec4440-III-DL.pdf.

Schuster, G. Spectrum estimation using Periodogram, Bartlett and Welch. Accessed June 24, 2016, http://www.laurent-duval.eu/Documents-Common/Schuster_G_2010_lect_spectrum_upbw.pdf.

(contain case studies [and some codes] of the Bartlett procedure along with several others [Welch, Blackman–Tukey, et al.] for improving the periodogram in detecting peaks in the spectrum of a noisy signal)

Jernigan, E. and Hui, E. Nonparametric power spectrum estimation methods. University of Waterloo, Waterloo, Ontario, Canada. Accessed June 24, 2016, http://miami.uwaterloo.ca/docs/SYDE770aReport.pdf.

(applies these techniques to the detection of prostate cancer)

3.9 OPTIONAL READING
THE SECRET OF BARTLETT'S METHOD

In estimating the autocorrelation and the power spectral density by applying the FFT, it seems surprising that simply incorporating more data into the FFT does not improve the results (while Bartlett's method of juggling the same data *does*). Throughout mathematics, we are conditioned to expect that "letting N go to infinity" effectuates accuracy and, ultimately, convergence.

To understand the wisdom of the Bartlett method, we first focus on a simpler problem in statistical estimation; later, we will show how the spectral estimation of autocorrelation falls into this framework.

Suppose A is a complex-valued random variable and we have several estimates $A_1, A_2, \ldots, A_n$ of A. The estimates come from data and are also (complex) random variables. We assume that

i. The estimates are independent, so that for $i \neq j$,

$$E\left\{A_i^p A_j^q\right\} = E\left\{A_i^p\right\} E\left\{A_j^q\right\}; \quad \text{also}$$

$$E\left\{A_i^p \overline{A}_j^q\right\} = E\left\{A_i^p\right\} E\left\{\overline{A}_j^q\right\}, E\left\{\overline{A}_i^p \overline{A}_j^q\right\} = E\left\{\overline{A}_i^p\right\} E\left\{\overline{A}_j^q\right\}. \tag{3.34}$$

ii. A and the estimates $A_1, A_2, \ldots, A_n$ are identically distributed, having identical moments. For simplicity, we take the mean to be zero and we call the variance U:

$$E\left\{A\right\} = E\left\{A_1\right\} = E\left\{A_2\right\} = \cdots = E\left\{A_n\right\} = 0. \tag{3.35}$$

$$E\left\{\left|A\right|^2\right\} = E\left\{\left|A_1\right|^2\right\} = E\left\{\left|A_2\right|^2\right\} = \cdots = E\left\{\left|A_n\right|^2\right\} = U. \tag{3.36}$$

Thus, we already know $E\{A\}$. We wish to form a good estimator for the squared magnitude $U = E\{|A|^2\}$, from the estimates $\{A_i\}$. Two procedures come to mind:

a. Magnitude-square the estimates and improve by taking the average:

$$\hat{U}_a = \frac{1}{n}\sum_{k=1}^{n}|A_k|^2. \tag{3.37}$$

b. Use averaging to obtain an improved estimate of A, and square the estimate:

$$\hat{U}_b = \left|\frac{1}{n}\sum_{k=1}^{n}A_k\right|^2. \tag{3.38}$$

Which is best?*

To level the playing field, we begin by checking that the estimators are unbiased, that is, they have the correct mean. For the first,

$$E\{\hat{U}_a\} = \frac{1}{n}\sum_{k=1}^{n}E\{|A_k|^2\} = U.$$

But the second estimator is biased and needs to be rescaled:

$$E\{\hat{U}_b\} = E\left\{\left|\frac{1}{n}\sum_{k=1}^{n}A_k\right|^2\right\} = E\left\{\left[\frac{1}{n}\sum_{k=1}^{n}A_k\right]\left[\frac{1}{n}\sum_{j=1}^{n}\bar{A}_j\right]\right\}$$

$$= \frac{1}{n^2}\sum_{k=1}^{n}\sum_{j=1}^{n}E\{A_k\bar{A}_j\} = \frac{1}{n^2}\sum_{k=1}^{n}U = \frac{nU}{n^2} = \frac{U}{n}.$$

So we replace the second estimator by

$$\hat{U}_c = n\hat{U}_b = n\left|\frac{1}{n}\sum_{k=1}^{n}A_k\right|^2 = \left|\frac{1}{\sqrt{n}}\sum_{k=1}^{n}A_k\right|^2. \tag{3.39}$$

Now we compare the estimators $\hat{U}_a$, $\hat{U}_c$ by examining the mean square errors, as discussed in Section 2.3. (Don't lose track; we are estimating the *square* of $|A|$, and we are assessing the *squared error* in the estimators. So expect to see some *fourth* powers coming into play.)

---

* Many of my students claim the answer is obvious and skip ahead to "The Bartlett Strategy."

For the first estimator,

$$MSE_a = E\left\{\left[\hat{U}_a - U\right]^2\right\} = E\{\hat{U}_a^2\} - U^2 \quad \text{(second moment identity)}$$

$$= E\left\{\left(\frac{1}{n}\sum_{k=1}^{n}|A_k|^2\right)\left(\frac{1}{n}\sum_{j=1}^{n}|A_j|^2\right)\right\} - U^2 = \frac{1}{n^2}\sum_{k=1}^{n}\sum_{j=1}^{n}E\left\{|A_k|^2|A_j|^2\right\} - U^2.$$

In the sum there are n terms where j and k match, and $n^2 - n$ terms where they differ. For the former, we have the fourth moment $E\{|A|^4\}$, and for the latter we have, by independence, $E\{|A|^2\}^2 = U^2$. Therefore,

$$MSE_a = \frac{1}{n^2}\left[nE\{|A|^4\} + (n^2 - n)U^2\right] - U^2 = \frac{1}{n}\left[E\{|A|^4\} - U^2\right].$$

We see that this estimator improves, and converges, as n gets larger.

For the (bias-corrected) second estimator,

$$MSE_c = E\left\{\left[\hat{U}_c - U\right]^2\right\} = E\{\hat{U}_c^2\} - U^2 \quad \text{(second-moment identity)}$$

$$= E\left\{\left(\frac{1}{\sqrt{n}}\sum_{k=1}^{n}A_k\right)\left(\frac{1}{\sqrt{n}}\sum_{k=1}^{n}\bar{A}_k\right)\left(\frac{1}{\sqrt{n}}\sum_{k=1}^{n}A_k\right)\left(\frac{1}{\sqrt{n}}\sum_{k=1}^{n}\bar{A}_k\right)\right\} - U^2$$

$$= \frac{1}{n^2}\sum_{j=1}^{n}\sum_{k=1}^{n}\sum_{l=1}^{n}\sum_{m=1}^{n}E\{A_j\bar{A}_kA_l\bar{A}_m\} - U^2.$$

In the sum we have the following groupings:

n terms where all four subscripts match yielding

$$E\left\{A_j\bar{A}_jA_j\bar{A}_j\right\} = E\left\{|A|^4\right\}$$

$\binom{4}{3}n(n-1)$ terms where three subscripts match, like $E\left\{A_j\bar{A}_jA_k\bar{A}_j\right\}$, yielding zero by independence: $E\left\{A_j\bar{A}_jA_k\bar{A}_j\right\} = E\left\{A_j\bar{A}_j\bar{A}_j\right\}E\left\{\bar{A}_k\right\} = E\left\{A_j\bar{A}_j\bar{A}_j\right\}\cdot 0$

$n(n-1)$ terms where two pairs of subscripts match* in the pattern $E\left\{A_j\bar{A}_jA_k\bar{A}_k\right\}$ yielding

$$E\left\{|A_j|^2\right\}E\left\{|A_k|^2\right\} = U^2$$

---

* The patterns where two pairs match are $E\left\{A_j\bar{A}_jA_k\bar{A}_k\right\}$, $E\left\{A_j\bar{A}_kA_j\bar{A}_k\right\}$, and $E\left\{A_j\bar{A}_kA_k\bar{A}_j\right\}$.

$n(n-1)$ terms where two pairs of subscripts match in the pattern $E\{A_j\bar{A}_k A_j\bar{A}_k\}$ yielding

$$E\{A_j^2\}E\{\bar{A}_k^2\} = E\{A^2\}E\{\bar{A}^2\} = \left|E\{A^2\}\right|^2$$

$n(n-1)$ terms where two pairs of subscripts match in the pattern $E\{A_j\bar{A}_k A_k\bar{A}_j\}$ yielding

$$E\{|A_j|^2\}E\{|A_k|^2\} = U^2$$

$\binom{4}{2}n(n-1)(n-2)$ terms where (only) one pair of subscripts matches, like $E\{A_j\bar{A}_j A_k\bar{A}_l\}$, yielding zero by independence:

$$E\{A_j\bar{A}_j A_k\bar{A}_l\} = E\{A_j\bar{A}_j\}E\{A_k\}E\{\bar{A}_l\} = E\{A_j\bar{A}_j\}\cdot 0\cdot 0$$

$n(n-1)(n-2)(n-3)$ terms where no subscripts match, like $E\{A_j\bar{A}_k A_l\bar{A}_m\}$, yielding zero by independence: $E\{A_j\bar{A}_k A_l\bar{A}_m\} = E\{A_j\}E\{\bar{A}_k\}E\{A_l\}E\{\bar{A}_m\} = 0\cdot 0\cdot 0\cdot 0$

Therefore, $MSE_c = \dfrac{1}{n^2}\left[nE\{|A|^4\} + n(n-1)E\{A^2\}^2 + 2n(n-1)U^2\right] - U^2$

$$= E\{A^2\}^2 + U^2 + \frac{E\{|A|^4\} - E\{A^2\}^2 - 2U^2}{n}$$

and error persists as n increases. This is not a good estimator.

The Bartlett Strategy: How does this analysis explain the superiority of the Bartlett approach? First, we need to characterize how the FFT works. Given M data points $X(n)$ the FFT returns, in essence (specific software implementations differ), the sums $\sum_{n=1}^{M} X(n)e^{-j2\pi nf}$ for M values of the frequency

$$f = \left\{-\frac{1}{2}, -\frac{1}{2}+\frac{1}{M}, -\frac{1}{2}+\frac{2}{M}, \cdots, \frac{1}{2}-\frac{1}{M}\right\} \text{ (cycles per sample)}.$$

Therefore, when the FFT is applied to a data set of, say, $2N = 32$ values, it returns transform estimates $\sum_{n=-16}^{15} X(n)e^{-j2\pi nf}$ for the 32 frequencies

$$-\frac{1}{2}, -\frac{1}{2}+\frac{1}{32}, -\frac{1}{2}+\frac{2}{32}, -\frac{1}{2}+\frac{3}{32}, -\frac{1}{2}+\frac{4}{32},$$

$$-\frac{1}{2}+\frac{5}{32}, \cdots, -\frac{1}{2}+\frac{8}{32}, \cdots, \frac{1}{2}-\frac{4}{32}, \cdots, \frac{1}{2}-\frac{1}{32}. \tag{3.40}$$

If we partition the data into four Bartlett blocks of eight points each, the FFT returns four transform estimates

$$\sum_{n=-16}^{-9} X(n)e^{-j2\pi nf}, \quad \sum_{n=-8}^{-1} X(n)e^{-j2\pi nf}, \quad \sum_{n=0}^{7} X(n)e^{-j2\pi nf}, \quad \sum_{n=8}^{15} X(n)e^{-j2\pi nf}$$

for each of the frequencies

$$-\frac{1}{2}, -\frac{1}{2}+\frac{1}{8}, -\frac{1}{2}+\frac{2}{8}, \cdots, \frac{1}{2}-\frac{1}{8}, \tag{3.41}$$

a subset of those in (3.40). But *at these frequencies, the 32-point FFT equals the sum of the four 8-point FFTs.* $\left(\right.$ Proof: If $f = -\dfrac{1}{2}+\dfrac{4r}{32} = -\dfrac{1}{2}+\dfrac{r}{8}$, then

$$\sum_{n=-16}^{15} X(n)e^{-j2\pi nf} = \sum_{n=-16}^{-9}\left[X(n)e^{-j2\pi nf} + X(n+8)e^{-j2\pi(n+8)f} + X(n+16)e^{-j2\pi(n+16)f}\right.$$

$$\left. + X(n+24)e^{-j2\pi(n+24)f}\right]$$

and

$$e^{-j2\pi(n+8)f} = e^{-j2\pi nf}e^{-j2\pi 8(-1/2+r/8)} = e^{-j2\pi nf}e^{-j2\pi(-4+r)} = e^{-j2\pi nf} = e^{-j2\pi(n+16)f} = e^{-j2\pi(n+24)f},$$

so

$$\sum_{n=-16}^{15} X(n)e^{-j2\pi nf}$$

$$= \sum_{n=-16}^{-9}\left[X(n)e^{-j2\pi nf} + X(n+8)e^{-j2\pi nf} + X(n+16)e^{-j2\pi nf} + X(n+24)e^{-j2\pi nf}\right]$$

$$= \sum_{n=-16}^{-9} X(n)e^{-j2\pi nf} + \sum_{n=-8}^{-1} X(n)e^{-j2\pi nf} + \sum_{n=0}^{7} X(n)e^{-j2\pi nf} + \sum_{n=8}^{15} X(n)e^{-j2\pi nf}.\Bigg)$$

Now, the squared magnitudes of each of the last four sums provide estimators for the power spectral density—as does the squared magnitude of the whole sum:

$$S_X(f) \approx P_N(f), \quad P_N(f) = \frac{\left| \sum_{\# \text{terms}} X(n)e^{-j2\pi nf} \right|^2}{\# \text{terms}}.$$

So is it better to average the squared magnitudes of each sum or to average the sums and then square it? The former is the Bartlett procedure and the latter is the customary procedure. Our analysis at the beginning of the section tells us that Bartlett is superior—as long as the conditions (i, ii) (3.34–3.36) hold.

Are the distributions identical? Since the random process must be stationary for us to even contemplate spectral analysis, each of the estimates will have the same distribution. And if the mean of each $X(n)$ is zero, the estimates are zero-mean also.

What about independence? This is trickier. The terms $\{X(n)\}$ *are* correlated, but if the autocorrelation $R_X(n)$ is negligible for, say, $n > K$, then an argument can be made that $X(n_1)$ and $X(n_2)$ are essentially independent for $| n_1 - n_2 | > K$ (and we say that K is the "correlation length"). So if the Bartlett data blocks are separated by the correlation length, the corresponding estimates can be taken as independent.

But this is not true, since the Bartlett blocks are adjacent; the $X(n)$ at the end of one block is right next to the $X(n)$ at the beginning of the next block. However, if the blocks are much longer than the correlation length, *most* of the data in the Bartlett-block sums are remote from, and thus independent of, the data in the other blocks.

Bottom line: the MSE in the Bartlett estimator for the power spectral density approaches zero as the number of blocks is increased, up to a certain point; the block sizes have to be longer than, say, two correlation lengths.

3.10   SPECTRAL ANALYSIS FOR CONTINUOUS RANDOM PROCESSES

Measured data—numerical data, certainly—come in discrete form "$X(n)$," and the practicum-based methods that we have studied in this section are premised on this assumption. In Chapter 4, however, we will consider physical *models* of random processes, and many of them will be parameterized by a time continuum—hence, "$X(t)$." So we pause to summarize the analogous spectral concepts and equations for dealing with (stationary) continuous processes.*

The natural connection between a discrete random process and a continuous process is the presumption that the discrete data are sampled values of the continuous data. To facilitate understanding we change notation and employ X_n for the discrete process, reserving $X(t)$ for the continuous process. The sampling effect is characterized by the equation

$$X_n = X(t) \qquad \text{when } t = n \, \Delta t, \tag{3.42}$$

---

* In designating $X(t)$ as a "continuous random process," we are only presuming that "t" parameterizes a continuum. We do not mean to imply that $X(t)$ is a continuous function nor that the range of X is a continuum.

with Δt denoting the spacing between samples and n running from minus infinity to plus infinity. The formulas for the spectral parameters for continuous processes are then derived from those for the discrete versions by taking limits as Δt approaches zero while enforcing (3.42) (and heeding the wisdom of Section 3.2) and imposing the requisite convergence assumptions. Skipping the details, we summarize the results.

SUMMARY: SPECTRAL PROPERTIES OF DISCRETE AND CONTINUOUS RANDOM PROCESSES

Discrete Process Concept	Continuous Process Concept
(Frequency f is measured in cycles per sample $-\frac{1}{2} < f < \frac{1}{2}$)	(Frequency f is measured in cycles per second $-\infty < f < \infty$)

Ergodicity of mean

$$E\{X_n\} = \lim_{N \to \infty} \frac{X_1 + X_2 + \cdots + X_N}{N} \qquad E\{X(t)\} = \lim_{T \to \infty} \frac{\int_{-T/2}^{T/2} X(\tau)\,d\tau}{T}$$

Ergodicity of autocorrelation

$$R_n = \lim_{N \to \infty} \frac{\sum_{1}^{N} X_m X_{m-n}}{N} \qquad R(t) = \lim_{T \to \infty} \frac{\int_{-T/2}^{T/2} X(\tau)X(\tau - t)\,d\tau}{T}$$

Power spectral density

$$S_X(f) = \sum_{k=-\infty}^{\infty} R_X(k)e^{-j2\pi kf} \qquad S_X(f) = \int_{-\infty}^{\infty} R_X(t)e^{-j2\pi tf}\,dt$$

$$R_X(n) = \int_{-1/2}^{1/2} S_X(f)e^{j2\pi nf}\,df \qquad R_X(t) = \int_{-\infty}^{\infty} S_X(f)e^{j2\pi tf}\,df$$

Linear system impulse response

$$Y(n) = \sum_{m=-\infty}^{\infty} h(n-m)X(m) \qquad Y(t) = \int_{-\infty}^{\infty} h(t-\tau)X(\tau)\,d\tau$$

$$S_Y(f) = \left|\tilde{H}(f)\right|^2 S_X(f) \qquad S_Y(f) = \left|\tilde{H}(f)\right|^2 S_X(f)$$

These relationships will be used in Chapter 4.

EXERCISES

Section 3.1

1. Consider the random processes described in the following list of problems from Chapter 2. Discuss the wide sense stationarity and ergodicity of each of the following problems:
   a. Problem 1 b. Problem 2 c. Problem 3 d. Problem 4
   e. Problem 5 f. Problem 6 g. Problem 7 h. Problem 8

     i. Problem 9   **j.** Problem 11   **k.** Problem 14   **l.** Problem 15
    m. Problem 18   **n.** Problem 19 **o.** Problem 20 **p.** Problem 22
    q. Problem 23   **r.** Problem 24 **s.** Problem 25

2. Discuss the wide sense stationarity and ergodicity of each of the following discrete processes:

  a. $X(n) = A$, where A is a random variable with probability density function $f_A(a)$.

  b. $X(n) = A\cos n\omega$, where A is $N(\mu, \sigma)$ and ω is constant.

  c. $X(n) = A\cos(n\omega + \Phi)$, where A and ω are constant and Φ is uniformly distributed between $-\pi$ and π.

  d. $X(n) = A\cos n\omega + B\sin n\omega$ where ω is constant and A and B are uncorrelated zero mean random variables with variance σ^2.

  e. A Bernoulli process where $X(n)$ takes the value 1 with probability 0.6 and 0 with probability 0.4.

3. *The autocorrelation as a* **positive definite function**. The autocorrelation of a stationary process, $R_X(t)$, has some interesting mathematical properties. Try to prove the following, or look them up in any of the advanced random process textbooks:

  a. $R_X(t)$ has a real, nonnegative Fourier transform $(S_X(f))$.

  b. $|R_X(t)|$ takes its maximum value at $t = 0$.

  c. $R_X(t) = R_X(-t)$.

  d. $R_X(t)$ is continuous at all t if it is continuous at $t = 0$.

  e. $R_X(t)$ is positive definite: for any a_i, a_j, t_i, t_j: $\sum_{i,j} a_i a_j R_X(t_i - t_j) \geq 0$.

    (Bochner's theorem* states that properties *a* and *e* are equivalent.)

    A table of Fourier transforms indicates that all of the following are valid autocorrelation functions (because of property a):

$$1, \delta(t), e^{-|t|}, e^{-t^2}, \cos t, \frac{1}{1+|t|}, \frac{1}{1+t^2}, \frac{\sin t}{t}, \frac{\sin^2 t}{t^2}, \begin{cases} 1 & \text{for } |t| < 1 \\ 0 & \text{for } |t| \geq 1 \end{cases}, \begin{cases} 1-|t| & \text{for } |t| < 1 \\ 0 & \text{for } |t| \geq 1 \end{cases}$$

4. The stationary random process $X(t)$ has autocorrelation $R_X(\tau)$ and (second-order) joint pdf $f_{X(t_1),X(t_2)}(x_1, x_2)$.

  a. Express the probability that $|X(t_1) - X(t_2)|$ is greater than a given positive number in terms of $f_{X(t_1),X(t_2)}(x_1, x_2)$.

  b. Apply Chebyshev's inequality (Problem 15 of Chapter 1) to the variable $X(t_1) - X(t_2)$ to show that this probability is less than or equal to $2[R(0) - R(\tau)]/a^2$.

---

* Bochner, S. 1932. *Vorlesungen über Fouriersche integrale*. Leipzig, Germany: Akademische Verlagsgesellschaft.

Section 3.7

5. Define the random variable $A = \dfrac{1}{2T} \displaystyle\int_{-T}^{T} \big[15 + v(t)\big]\,dt$ where $v(t)$ is white noise with power 5. What are the mean and variance of A?

6. If white noise with autocorrelation function $R_W(\tau) = \delta(\tau)$ is passed through a filter with unit impulse response function $h(t)$, prove that the output $Y(t)$ has average power

$$E\{Y^2(t)\} = \int_{-\infty}^{\infty} h^2(t)\,dt.$$

7. **a.** What is the unit impulse voltage response of the RC circuit in Figure 3.11 to a current input?

   b. If the input current $X(t)$ has zero mean and autocorrelation $R_X(\tau) = 5e^{-|\tau|}$ and the output voltage $Y(t)$ has average power 10, find the output autocorrelation $R_Y(\tau)$ and the value of RC.

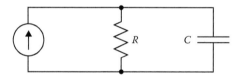

FIGURE 3.11   RC circuit.

8. For the stationary autoregressive process modeled by

$$X(n) = 0.3X(n-1) + 0.25X(n-2) + 3V(n),$$

where $V(n)$ is white noise, independent of X, with power $\sigma_V^2 = 2$, what is the power spectral density of the process $X(n)$?

Section 3.8

9. Create a MATLAB™ M-file containing the following commands, and call it "Bartlett.m." When you type Bartlett in the MATLAB window and press Enter successively, you will see the following (compare Figures 3.6 through 3.9):

   i. A simulated 1024-point random process (ARMA(2,0), to be described in Section 4.3) with autocorrelation $R_X(n) = (-a)^n/[1 - a^2]$, where a is a parameter that you will provide in the range $-1 < a < 1$

   ii. The periodogram of the process superposed over the theoretical power spectral density

    iii. The Bartlett estimate of the power spectral density obtained by averaging
        over 4 periodogram blocks of 256 points each
    iv. The Bartlett estimate of the power spectral density obtained by averaging
        over 16 periodogram blocks of 64 points each
    v. The previous three graphs, superposed
    vi. The autocorrelation estimated from the 16-block average
Experiment with the values $a = -0.8, -0.2, 0, 0.2, 0.8$; explain what you are seeing.

_____Bartlett.m_____

```
Pi=3.1415927;
randn('seed',0);
y=randn(1024,1);
b=[1];
a=input('Autocorrelation parameter -1 < a < 1? ')
a=[1.0 a];
x=filter(b,a,y);
X=fft(x,1024);
P=X.*conj(X)/1024;
figure;
plot(x)
AA=num2str(a); BB=num2str(b);
heading=['ARMA signal: a = [', AA, '], b = [', BB, ']'];
title(heading);
pause;
figure
P=fftshift(P);
x1024=linspace(0,1023,1024);
% axis([0 1024 0 12])
[H,ww]=freqz(b,a,1024,'whole');
H=fftshift(H);
plot(x1024,P,'m.-',x1024,abs(H.^2),'g.-')
YMAX=max(P);
axis([0 1024 0 YMAX])
title('Periodogram of x[n] with N=1024');
pause;
x1=x(1:256);
x2=x(257:512);
x3=x(513:768);
x4=x(769:1024);
X1=fft(x1,256);
X2=fft(x2,256);
X3=fft(x3,256);
X4=fft(x4,256);
P1=X1.*conj(X1)/256;
P2=X2.*conj(X2)/256;
P3=X3.*conj(X3)/256;
P4=X4.*conj(X4)/256;
B=(P1+P2+P3+P4)/4;
figure
B4=fftshift(B);
```

```
x256=linspace(1,255,256);
% plot(x1024,P,'m.-',x256,B4,'r.-')
[H,ww]=freqz(b,a,256,'whole');
H=fftshift(H);
plot(x256,B4,'r.-',x256,abs(H.^2),'g.-')
axis([0 256 0 YMAX]);
title('Bartlett Estimate with 4 averaged periodogram blocks')
pause;
x1=x(1:64);
x2=x(65:128);
x3=x(129:192);
x4=x(193:256);
x5=x(257:320);
x6=x(321:384);
x7=x(385:448);
x8=x(449:512);
x9=x(513:576);
x10=x(577:640);
x11=x(641:704);
x12=x(705:768);
x13=x(769:832);
x14=x(833:896);
x15=x(897:960);
x16=x(961:1024);
X1=fft(x1,64);
X2=fft(x2,64);
X3=fft(x3,64);
X4=fft(x4,64);
X5=fft(x5,64);
X6=fft(x6,64);
X7=fft(x7,64);
X8=fft(x8,64);
X9=fft(x9,64);
X10=fft(x10,64);
X11=fft(x11,64);
X12=fft(x12,64);
X13=fft(x13,64);
X14=fft(x14,64);
X15=fft(x15,64);
X16=fft(x16,64);
P1=X1.*conj(X1)/64;
P2=X2.*conj(X2)/64;
P3=X3.*conj(X3)/64;
P4=X4.*conj(X4)/64;
P5=X5.*conj(X5)/64;
P6=X6.*conj(X6)/64;
P7=X7.*conj(X7)/64;
P8=X8.*conj(X8)/64;
P9=X9.*conj(X9)/64;
P10=X10.*conj(X10)/64;
P11=X11.*conj(X11)/64;
```

```
P12=X12.*conj(X12)/64;
P13=X13.*conj(X13)/64;
P14=X14.*conj(X14)/64;
P15=X15.*conj(X15)/64;
P16=X16.*conj(X16)/64;
B=(P1+P2+P3+P4+P5+P6+P7+P8+P9+P10+P11+P12+P13+P14+P15+P16)/16;
figure
B16=fftshift(B);
x64=linspace(0,63,64);
[H,ww]=freqz(b,a,64,'whole');
H=fftshift(H);
plot(x64,B16,'k.-',x64,abs(H.^2),'g.-')
axis([0 64 0 YMAX]);
title('Bartlett Estimate with 16 averaged periodogram
blocks');
pause;
figure
plot(x1024,P,'m.-',x256,B4,'r.-',x64,B16,'k.-')
title('Summary of periodogram plots')
pause;
C=real(ifft(B));
xc=linspace(0,31,32);
figure
plot(xc,C(1:32));
title('Autocorrelation');
```

4 Models for Random Processes

Now, we turn to the analysis of random processes for which we have theoretical models—rather than simply data. Therefore, the procedures for deriving such parameters as needed for prediction (such as the autocorrelation or power spectral density [PSD]) will now be theory based. Of course, if a set of raw data generates an autocorrelation similar to one derived in this chapter, we'll have a model for the data.

4.1 DIFFERENTIAL EQUATIONS BACKGROUND

In order to get some experience in simulation and testing of random processes, it is helpful to have a simple tool that will generate random processes with specified properties, such as mean and autocorrelation. Such a tool is provided by difference equation models. For the sake of exposition, we shall describe this model by starting with a more familiar one, the differential equation. When we are done, we shall see how to construct a difference equation that, together with a basic random number generator, generates random processes having prescribed autocorrelations. We can use this then for generating random processes with prescribed properties for simulation and constructing a feasible model for a specific (measured) random process whose means and autocorrelations are obtained as in Chapter 3.

Recall the basics of the theory of linear nonhomogeneous constant-coefficient ordinary differential equations* such as

$$\frac{d^2x}{dt^2} + 2\frac{dx}{dt} + 2x = 5\sin t, \quad x(0) = 1, \quad \frac{dx(0)}{dt} = 2. \tag{4.1}$$

Typically, such equations are solved using the following steps:

1. Try to find a *particular* solution that closely matches the form of the non-homogeneity. In this case, $x_{part}(t) = A\sin t + B\cos t$ is a near match to the nonhomogeneity $5\sin t$, and trial-and-error (the "method of undetermined coefficients") reveals a solution to be $x_{part}(t) = \sin t - 2\cos t$.
2. Find the general solution to the *homogeneous* equation associated with (4.1),

$$\frac{d^2x}{dt^2} + 2\frac{dx}{dt} + 2x = 0, \tag{4.2}$$

---

* Your author's favorite textbook on differential equations is Nagle, R.K., Saff, E.B., and Snider, A.D. 2016. *Fundamentals of Differential Equations and Boundary Value Problems*, 7th edn. Boston, MA: Pearson.

by substituting the "trial solution" $x(t)=e^{\lambda t}$. Here the suitable values for λ are found to be $-1 \pm i$, yielding the general solution

$$x_{homog}(t)=c_1 e^{-t}\cos t+c_2 e^{-t}\sin t.$$

3. Choose values for c_1 and c_2 that enforce the initial conditions for the complete solution:

$$x(t)=x_{part}(t)+x_{homog}(t)= \sin t-2\cos t+c_1 e^{-t}\cos t+c_2 e^{-t}\sin t.$$

By substituting in (4.1), we find $c_1 = 3$, $c_2 = 4$.

Although (4.1) is by no means the most general differential equation of its type, its solution $x(t) = \sin t - 2 \cos t + 3e^{-t} \cos t + 4e^{-t}\sin t$ has a feature that is typical of most engineering system responses, namely, there are terms that decay as $t \rightarrow \infty$ (those that contain e^{-t}), leaving a persisting "steady-state" response $x_{steady\,state}(t) = \sin t - 2 \cos t$. The latter solution is independent of the initial conditions; in fact, *every* solution of the differential equation tends to the steady state solution. (Why? Because of the form of the *general* solution expression:

$$\sin t - 2\cos t + c_1 e^{-t}\cos t + c_2 e^{-t}\sin t.)$$

4.2 DIFFERENCE EQUATIONS

One well-known method of obtaining approximate numerical solutions to a differential equation is to replace the unknown derivatives therein by their *finite-difference* approximations. Recall the mathematical definition of the derivative:

$$\frac{dx}{dt} = \lim_{\Delta t \to 0} \frac{x(t+\Delta t)-x(t)}{\Delta t}. \tag{4.3}$$

We exploit this by proposing the approximation

$$\frac{dx}{dt} \approx \frac{x(t+\Delta t)-x(t)}{\Delta t} \tag{4.4}$$

for small Δt. A limit expression for the second derivative can be derived from (4.4) by iterating the same logic:

$$\frac{d^2 x}{dt^2} = \lim_{\Delta t' \to 0} \frac{\frac{dx}{dt}(t+\Delta t')-\frac{dx}{dt}(t)}{\Delta t'}$$

$$= \lim_{\Delta t' \to 0} \frac{\lim_{\Delta t \to 0}\frac{x(t+\Delta t'+\Delta t)-x(t+\Delta t')}{\Delta t}-\lim_{\Delta t \to 0}\frac{x(t+\Delta t)-x(t)}{\Delta t}}{\Delta t'}$$

$$= \lim_{\Delta t' \to 0}\lim_{\Delta t \to 0}\frac{x(t+\Delta t'+\Delta t)-x(t+\Delta t')-x(t+\Delta t)+x(t)}{\Delta t'\Delta t}.$$

Taking $\Delta t' \equiv \Delta t$ for simplicity, we propose the approximation

$$\frac{d^2x}{dt^2} \approx \frac{x(t+2\Delta t) - 2x(t+\Delta t) + x(t)}{\Delta t^2}. \tag{4.5}$$

Skilled numerical analysts can develop better approximations than these, but (4.4) and (4.5) will suffice to motivate our deliberations.

If we were to try to approximate the values of the solution $x(t)$ to the differential equation 4.1 of the preceding section on a discrete mesh of time values $t_j = j\Delta t$ ($j = 0$, 1, 2, ...), we would replace the derivatives therein by their approximations:

$$\frac{x(t+2\Delta t) - 2x(t+\Delta t) + x(t)}{\Delta t^2} + 2\frac{x(t+\Delta t) - x(t)}{\Delta t}$$

$$+ 2x(t) = 5\sin t, \quad x(0) = 1, \quad \frac{x(0+\Delta t) - x(0)}{\Delta t} = 2. \tag{4.6}$$

We then have a *finite-difference equation* approximating the original system. If we rearrange Equation 4.6

$$x(0) = 1$$
$$x(\Delta t) = 2\Delta t + x(0) = 2\Delta t + 1$$
$$x(t+2\Delta t) = \{2 - 2\Delta t\} x(t+\Delta t) + \{-1 + 2\Delta t - 2\Delta t^2\} x(t) - 5\Delta t^2 \sin t,$$

we get a recursive scheme for easily computing $x(n\Delta t)$, ... in terms of the previous two values $x([n-1]\Delta t)$ and $x(n\Delta t)$ and of the nonhomogeneity $\sin(n\Delta t)$:

$$x(0) = 1$$
$$x(\Delta t) = 2\Delta t + 1$$
$$x(2\Delta t) = \{2 - 2\Delta t\} x(\Delta t) + \{-1 + 2\Delta t - 2\Delta t^2\} x(0) - 5\Delta t^2 \sin 0$$
$$x(3\Delta t) = \{2 - 2\Delta t\} x(2\Delta t) + \{-1 + 2\Delta t - 2\Delta t^2\} x(\Delta t) - 5\Delta t^2 \sin \Delta t$$
$$\vdots$$
$$x(n\Delta t) = \{2 - 2\Delta t\} x([n-1]\Delta t) + \{-1 + 2\Delta t - 2\Delta t^2\} x([n-2]\Delta t) - 5\Delta t^2 \sin(n-2)\Delta t.$$

More generally, we call any equation (whether or not it was derived from a differential equation) of the form

$$x(n) = a(1)x(n-1) + a(2)x(n-2) + \cdots + a(p)x(n-p)$$
$$+ b(0)v(n) + b(1)v(n-1) + \cdots + b(q)v(n-q) \tag{4.7}$$

a linear, constant-coefficient, pth-order difference equation for the (unknown) sequence $\{x(n)\}$, driven by the (known) nonhomogeneous sequence $\{v(n)\}$. The steps for solving such a difference equation analytically are very similar to what we outlined in Section 4.1 for differential equations:

1. Try to find a *particular* solution $x_{part}(n)$ by exploiting the form of the nonhomogeneity $v(n)$.
2. Find a general solution to the associated homogeneous difference equation,

$$x(n) = a(1)x(n-1) + a(2)x(n-2) + \cdots + a(n-p)x(n-p), \qquad (4.8)$$

by substituting the trial solution $x(n) = e^{\lambda n} \equiv r^n$. If you carry this out, you will easily see that this results in a polynomial equation for r of degree p:

$$r^p - a(1)r^{p-1} - a(2)r^{p-2} - \cdots - a(p) = 0. \qquad (4.9)$$

As a rule, such an equation has p roots $\{r_1, r_2, \ldots, r_p\}$ and the general solution of (4.8) takes the form

$$x(n) = c_1 r_1^n + c_2 r_2^n + \cdots + c_p r_p^n,$$

with p arbitrary constants $\{c_i\}$. If the roots $\{r_i\}$ are repeating or complex, suitable adjustments can be made to produce p real, independent solutions, as in the case of constant-coefficient differential equations.
3. Assemble the general solution to the original equation

$$x(n) = x_{part}(n) + \sum_{i=1}^{p} c_i r_i^n \qquad (4.10)$$

and choose values for the $\{c_i\}$ that enforce the "initial conditions":

$$x(0) = x_0, \quad x(1) = x_1, \ldots, x(p-1) = x_{p-1}.$$

Clearly from (4.10), the difference equation is "unstable," in the sense that it might have unbounded solutions if any of the $\{r_i\}$ has magnitude greater than one. On the other hand if each $\{r_i\}$ has magnitude less than one, the terms $\sum_{i=1}^{p} c_i r_i^n$ become arbitrarily small as n gets large, and the solution approaches a "steady-state" form that is independent of the starting values.*

---

* Using Rouche's Theorem (see any complex variables text; your author's personal favorite is Saff, E.B. and Snider, A.D. 2003. *Fundamentals of Complex Analysis*, 3rd edn. Boston, MA: Pearson), one can prove that stability is guaranteed if the coefficients satisfy $\sum_{i=1}^{p} |a(i)| < 1$.

4.3   ARMA MODELS

As a special case of the general difference equations (4.7), Section 4.2, take a look at the form of the difference equation if all $a(i)$ are zero and each $b(i)$ equals $1/(q + 1)$:

$$x(n) = b(0)v(n) + b(1)v(n-1) + \cdots + b(q)v(n-q) = \frac{\sum_{i=0}^{q} v(n-i)}{q+1}.$$

In this case, $x(n)$ merely averages the last $(q + 1)$ values of the *(known)* *nonhomogeneity sequence* $\{v(n)\}$. This motivates the terminology *moving average* for the nonhomogeneous terms in the difference equation. Similarly, the "feedback" aspect suggested by the associated homogeneous equation (4.8), Section 4.2,

$$x(n) = a(1)x(n-1) + a(2)x(n-2) + \cdots + a(p)x(n-p), \qquad (4.11)$$

has led to the jargon *autoregression* for these terms. Thus, the nomenclature **autoregressive moving average**, or ARMA(p, q), is used in the statistics literature to describe such difference equations:

$$x(n) = \underbrace{a(1)x(n-1) + a(2)x(n-2) + \cdots + a(p)x(n-p)}_{\text{autoregressive of order } p}$$
$$\underbrace{+ b(0)v(n) + b(1)v(n-1) + \cdots + b(q)v(n-q)}_{\text{moving average of order } q}. \qquad (4.12)$$

For the purposes of modeling random processes, the nonhomogeneity sequence in the ARMA equation is taken to be a sequence of random variables that we distinguish by capitalization: $V(n)$. The $V(n)$ are identically distributed and independent of each other. Each has mean 0 and standard deviation σ.

$$E\{V(n)\} = 0; \quad E\{V(n)^2\} = \sigma^2 = 0; \quad E\{V(n)V(m)\} = 0 \quad for \ m \neq n. \ (4.13)$$

The sequence $V(n)$, thus, has the characteristics of white noise (Section 3.6), and for simulations they can be generated by a random number generator. The associated ARMA sequence generated, $X(n)$, is random also, but because of the coupling-in of earlier values of X and V in Equation 4.12, the values of $X(n)$ will be correlated. We shall investigate this correlation shortly.

The particular sequence $\{X(n)\}$ generated by an ARMA equation (4.12) depends on the choice of starting values $X(1)$, $X(2)$, ..., $X(p)$. However, if the equation is stable, that is, all the roots of Equation 4.9 are less than one in magnitude, then all of the sequences tend (as $n \to \infty$) to a common limiting sequence determined only by the values of the $\{V(n)\}$.* This common limiting sequence,

---

* Mathematical argument: Any two sequences generated by a common set of $\{V(n)\}$ differ by a solution to the associated homogeneous autoregressive equation (4.8); and all such solutions approach zero.

which we can obtain in practice by iterating (4.12) until the "transients" die out, is then stationary and has mean zero.

4.4 THE YULE–WALKER EQUATIONS

Under the presumption that the sequence $X(n)$ is stationary, the *Yule–Walker*[*,†] *equations* express the relation between its autocorrelations and the coefficients in the ARMA model. To simplify the derivation, we first restrict ourselves to the ARMA(2, 0) model:

$$\text{ARMA}(2,0): \quad X(n) = a(1)X(n-1) + a(2)X(n-2) + b(0)V(n). \quad (4.14)$$

In particular, write this equation for $n = 2$:

$$X(2) = a(1)X(1) + a(2)X(0) + b(0)V(2). \quad (4.15)$$

The "trick" for converting the ARMA equation (4.15) into a Yule–Walker relation for the autocorrelations is to multiply it $X(0)$, $X(1)$, $X(2)$, and $V(2)$ in turn, and take expected values. Observe the following:

First, multiply (4.15) by $X(0)$ and take expected values:

$$E\{X(0)X(2)\} = a(1)E\{X(0)X(1)\} + a(2)E\{X(0)X(0)\} + b(0)E\{X(0)V(2)\}. \quad (4.16)$$

Note the occurrence of the autocorrelations $R_X(k) \equiv E\{X(m)\,X(m+k)\}$ for $k = 2$, 1, and 0. Note also that $V(2)$ only influences $X(2)$, $X(3)$, and $X(n)$ for $n \geq 2$; it is independent of $X(0)$ and of $X(1)$ (because they occur earlier in the sequence), and therefore, $E\{X(0)\,V(2)\} = E\{X(0)\}\,E\{V(2)\} = 0$. Thus, Equation 4.16 says

$$R_X(2) = a(1)R_X(1) + a(2)R_X(0). \quad (4.17)$$

Next, multiply (4.15) by $X(1)$ and take means:

$$E\{X(1)X(2)\} = a(1)E\{X(1)X(1)\} + a(2)E\{X(1)X(0)\} + b(0)E\{X(1)V(2)\}$$

or

$$R_X(1) = a(1)R_X(0) + a(2)R_X(1). \quad (4.18)$$

---

[*] Yule, G.U. 1927. On a method of investigating periodicities in disturbed series, with special reference to Wolfer's Sunspot numbers. *Philosophical Transactions of the Royal Society of London, Series A* 226: 267–298.

[†] Walker, G. 1931. On periodicity in series of related terms. *Proceedings of the Royal Society of London, Series A* 131: 518–532.

Multiplication of (4.15) by $X(2)$ and taking means yields

$$E\{X(2)X(2)\}=a(1)E\{X(2)X(1)\}+a(2)E\{X(2)X(0)\}+b(0)E\{X(2)V(2)\}$$

or

$$R_X(0)=a(1)R_X(1)+a(2)R_X(2)+b(0)E\{X(2)V(2)\}. \qquad (4.19)$$

The term $E\{X(2)V(2)\}$ is troublesome because $V(2)$ is *not* independent of $X(2)$; it occurs in Equation 4.15 for $X(2)$. However, a final multiplication of (4.15) by $V(2)$ and taking expected values gives us this term since

$$E\{V(2)X(2)\}=a(1)E\{V(2)X(1)\}+a(2)E\{V(2)X(0)\}+b(0)E\{V(2)V(2)\}$$

or

$$E\{V(2)X(2)\}=0+0+b(0)\sigma^2. \qquad (4.20)$$

Thus (4.19) reduces to

$$R_X(0)=a(1)R_X(1)+a(2)R_X(2)+b(0)^2\sigma^2. \qquad (4.21)$$

Equations 4.17, 4.18, and 4.21 constitute three coupled linear equations for $R_X(0)$, $R_X(1)$, and $R_X(2)$:

$$R_X(2)=a(1)R_X(1)+a(2)R_X(0).$$

$$R_X(1)=a(1)R_X(0)+a(2)R_X(1).$$

$$R_X(0)=a(1)R_X(1)+a(2)R_X(2)+b(0)^2\sigma^2. \qquad (4.22)$$

And they can be recast in matrix form and solved by standard methods:

$$\begin{bmatrix} -a(2) & -a(1) & 1 \\ -a(1) & [1-a(2)] & 0 \\ 1 & -a(1) & -a(2) \end{bmatrix} \begin{bmatrix} R_X(0) \\ R_X(1) \\ R_X(2) \end{bmatrix} = \begin{bmatrix} 0 \\ 0 \\ b(0)^2\sigma^2 \end{bmatrix}.$$

The calculation of the remaining values of the autocorrelation R_X is easy: multiplication of (4.14) by $X(0)$ for $n>2$ yields a simple recursion formula:

$$R_X(n)=a(1)R_X(n-1)+a(2)R_X(n-2) \quad (n>2). \qquad (4.23)$$

Therefore, we can easily calculate $R_X(3)$, $R_X(4)$, ... once we have found $R_X(2)$, $R_X(1)$, and $R_X(0)$ from (4.22).

The set of Equations 4.22 and 4.23 are the **Yule–Walker equation** for the ARMA(2, 0) model. They relate the autocorrelations to the model coefficients.

If the starting value $X(0)$ for ARMA(1, 0) is a zero-mean Gaussian random variable and the noise $V(n)$ is Gaussian, the resulting ARMA sequence is known as the **Ornstein–Uhlenbeck** process. It, too, is Gaussian (because sums of Gaussian random variables are also Gaussian—see Section 1.13).

ONLINE SOURCES

These simulators let you experiment with different ARMA models, displaying samples, computing and estimating autocorrelations, and even demonstrating (spectacularly) instability when the coefficients are too big:

Baniar, M. Autoregressive moving-average generator. Wolfram Demonstrations Project. Accessed June 24, 2016. http://demonstrations.wolfram.com/AutoregressiveMoving AverageGenerator/.
von Seggern, D. Auto-regressive simulation (second-order). Wolfram Demonstrations Project. Accessed June 24, 2016. http://demonstrations.wolfram.com/AutoRegressive SimulationSecondOrder/.

4.5   CONSTRUCTION OF ARMA MODELS

In the previous section we saw how to derive all the autocorrelations of a random process $X(n)$ from an ARMA(2, 0) model of the process. Now, we reverse our objective and attempt to construct the *model*, that is, the coefficients $a(-)$, $b(-)$, from the autocorrelation.*

A typical situation where such a calculation would be applicable occurs in **system identification**. Suppose one has a general picture of the mechanisms of some phenomena. (For example, a system of difference equations can be formulated for the workings of the human vocal system.) However, the values of the parameters (resistances/capacitances/ inductances, masses/damping/spring constants, …) are unavailable. One could drive the actual system with some random process having known characteristics—white noise, perhaps. Using Bartlett's method, one could then estimate the autocorrelation of the output. If an ARMA model is reconstructed from the autocorrelation, we could identify its coefficients with those in the difference equation model.

So, now, we assume we are *given* the autocorrelations of a stationary random process $X(n)$ and we seek to simulate it with, to be specific, an ARMA(2,0) model. We can write the first three Yule–Walker equations in (4.22), Section 4.4, to display the *coefficients* $a(2)$, $a(1)$, and $b(0)$ as unknowns:

$$\begin{bmatrix} R_X(1) & R_X(0) & 0 \\ R_X(0) & R_X(1) & 0 \\ R_X(1) & R_X(2) & \sigma^2 \end{bmatrix} \begin{bmatrix} a(1) \\ a(2) \\ b(0)^2 \end{bmatrix} = \begin{bmatrix} R_X(2) \\ R_X(1) \\ R_X(0) \end{bmatrix}. \tag{4.24}$$

---

* This is very similar to the task we described in Section 3.7, designing a filter to shape a PSD since the autocorrelation is the inverse discrete time Fourier transform of the PSD.

Note that Equations 4.24 form a *linear* system determining $a(2)$, $a(1)$, and $b(0)^2$; we need to take a square root to get $b(0)$. (Either root, plus or minus, is equally valid, because the postulated characteristics of the random process $V(n)$ (listed in (4.13)) are equally valid for the process—$V(n)$.)

So the calculated solution $a(1)$, $a(2)$, and $b(0)$ of (4.24) is consistent with the values of $R_X(0)$, $R_X(1)$, and $R_X(2)$. But what about the remaining Yule–Walker equations containing $R_X(3)$, $R_X(4)$, etc.? Will they be valid?

If the process $X(n)$ is *truly* an ARMA(2, 0) process, then all the Yule–Walker equations must be valid—and since the first subset (4.24) *determines* the coefficients *uniquely*, the remaining equations will automatically hold.

On the other hand, if the remaining equations do *not* hold, then we must conclude that $X(n)$ is not an ARMA(2, 0) process. We might try to fit an ARMA(3, 0) or ARMA(2, 1) process, and so on, using the higher-order Yule–Walker equations to be derived in Section 4.6. Of course, there is the possibility that $X(n)$ is not an ARMA process at all, and one would have to make a judgment as to how many autocorrelation "matches" will be good enough for the task at hand.

Now that we have mastered the ARMA(2, 0) process "forward and backward," we will have little trouble generalizing to the higher-order cases.

4.6   HIGHER-ORDER ARMA PROCESSES

To get a feeling for the Yule–Walker equations for a higher-order ARMA process, we shall work out the ARMA(3, 0) and ARMA(3, 1) cases. The procedure for ARMA(p, q) will then become clear.

$$\text{ARMA}(3,0): X(n) = a(1) X(n-1) + a(2) X(n-2) + a(3) X(n-3) + b(0) V(n). \tag{4.25}$$

Proceeding analogously to Section 4.4, we first express the difference equation for $X(3)$,

$$X(3) = a(1) X(2) + a(2) X(1) + a(3) X(0) + b(0) V(3), \tag{4.26}$$

and take expected values of products of (4.26) with $X(0)$, $X(1)$, $X(2)$, $X(3)$, and $V(3)$ in turn to derive equations for the autocorrelations:

$$R_X(3) = a(1) R_X(2) + a(2) R_X(1) + a(3) R_X(0). \tag{4.27}$$

$$R_X(2) = a(1) R_X(1) + a(2) R_X(0) + a(3) R_X(1). \tag{4.28}$$

$$R_X(1) = a(1) R_X(0) + a(2) R_X(1) + a(3) R_X(2). \tag{4.29}$$

$$R_X(0) = a(1) R_X(1) + a(2) R_X(2) + a(3) R_X(3) + b(0) E\{X(3) V(3)\}. \tag{4.30}$$

$$E\{X(3)V(3)\} = b(0)\sigma^2. \tag{4.31}$$

Substitution of (4.31) into (4.30) yields four linear Yule–Walker equations (4.27 through 4.30), determining $R_X(0)$, $R_X(1)$, $R_X(2)$, and $R_X(3)$.

Taking expected values of the product of $X(0)$ with (4.25) for $n > 3$ then yields the easily implemented recursive Yule–Walker equation for $n > 3$:

$$R_X(n) = a(1)R_X(n-1) + a(2)R_X(n-2) + a(3)R_X(n-3) \qquad (n > 3). \tag{4.32}$$

Conversely, if the autocorrelations are known, Equations 4.27 through 4.31 determine $a(1)$, $a(2)$, $a(3)$, and $|b(0)|$; the remaining equations in (4.32) confirm or deny the validity of the proposed ARMA(3, 0) model.

We can see that the calculation of *either* the autocorrelations or the coefficients for any autoregressive process ARMA(p, 0) (or AR(p), as it is sometimes called) is straightforward.*

So let's turn to the ARMA(p, q) process with $q > 0$. To be specific, we consider ARMA(3, 1):

$$X(n) = a(1)X(n-1) + a(2)X(n-2) + a(3)X(n-3) + b(0)V(n) + b(1)V(n-1). \tag{4.33}$$

First, we express the difference equation at the level $n = p + q = 3 + 1 = 4$,

$$X(4) = a(1)X(3) + a(2)X(2) + a(3)X(1) + b(0)V(4) + b(1)V(3), \tag{4.34}$$

and take expected values of the product of (4.34) with $X(0)$, $X(1)$, $X(2)$, $X(3)$, $X(4)$, $V(4)$, and $V(3)$. Bearing in mind that $X(m)$ is independent of $V(n)$ only if $m < n$, we find

$$R_X(4) = a(1)R_X(3) + a(2)R_X(2) + a(3)R_X(1). \tag{4.35}$$

$$R_X(3) = a(1)R_X(2) + a(2)R_X(1) + a(3)R_X(0). \tag{4.36}$$

$$R_X(2) = a(1)R_X(1) + a(2)R_X(0) + a(3)R_X(1). \tag{4.37}$$

$$R_X(1) = a(1)R_X(0) + a(2)R_X(1) + a(3)R_X(2) + b(1)E\{X(3)V(3)\}. \tag{4.38}$$

---

* The autoregressive process with $b(0) = 0$, depicted in Equation 4.11, is of no interest since the stationary steady state solution is identically zero.

$$R_X(0) = a(1)R_X(1) + a(2)R_X(2) + a(3)R_X(3)$$

$$+ b(0)E\{X(4)V(4)\} + b(1)E\{X(4)V(3)\}. \tag{4.39}$$

Since $E\{X(4)\,V(4)\} = (0) + (0) + (0) + b(0)\sigma^2 + (0) = b(0)\sigma^2$, it follows by stationarity that

$$E\{X(j)V(j)\} = b(0)\sigma^2 \qquad \text{for all } j, \tag{4.40}$$

and thus,

$$E\{X(4)\,V(3)\} = a(1)E\{X(3)V(3)\} + (0) + (0) + (0) + b(1)\,\sigma^2$$

$$= a(1)\,b(0)\,\sigma^2 + b(1)\sigma^2. \tag{4.41}$$

We substitute (4.40) and (4.41) into (4.38) and (4.39) and combine with (4.35) through (4.37) to obtain the Yule–Walker relations for $R_X(0)$ to $R_X(4)$:

$$\begin{bmatrix} 0 & -a(3) & -a(2) & -a(1) & 1 \\ -a(3) & -a(2) & -a(1) & 1 & 0 \\ -a(2) & [-a(3)-a(1)] & 1 & 0 & 0 \\ -a(1) & [1-a(2)] & -a(3) & 0 & 0 \\ 1 & -a(1) & -a(2) & -a(3) & 0 \end{bmatrix} \begin{bmatrix} R_X(0) \\ R_X(1) \\ R_X(2) \\ R_X(3) \\ R_X(4) \end{bmatrix}$$

$$= \begin{bmatrix} 0 \\ 0 \\ 0 \\ b(1)b(0)\sigma^2 \\ \{b(0)^2 + b(1)[a(1)b(0) + b(1)]\}\sigma^2 \end{bmatrix}. \tag{4.42}$$

For $n > 4$ the expected values of $X(0)$ times (4.34) yields the recursion for the remaining autocorrelations:

$$R_X(n) = a(1)R_X(n-1) + a(2)R_X(n-2) + a(3)R_X(n-3) \qquad (n > 4). \tag{4.43}$$

Thus the calculations of the *autocorrelations* for ARMA(3,1) is straightforward. In fact, following a similar pattern of reasoning (4.34 through 4.41) will always yield $(p + q + 1)$ linear equations for the first $p + q + 1$ autocorrelations of an ARMA(p, q) model and a recursion like (4.43) for the higher autocorrelations.

However, if we try to find *coefficients* to fit an ARMA(3, 1) to a *given* set of autocorrelations, the solution of Equations 4.42 for the coefficients $a(1)$, $a(2)$,

$a(3)$, $b(0)$, $b(1)$ is burdensome because the system is *non*linear (e.g., observe the term $b(1)a(1)b(0)$).

Thus the fitting of full ARMA(p, q) models with $q > 0$ to a given set of correlations is more complex that fitting AR models, and this inhibits the use of the former in some situations.*

4.7 THE RANDOM SINE WAVE

In Section 2.1, we considered the output of an AC signal generator as a random process by presuming that a saboteur had set the control dials to some unspecified values. Figure 2.8 in that section depicted sample waveforms when the amplitude was randomized, and Figure 2.9 depicted those with random phases. We might as well consider his tampering with the *frequency* dial as well. Then every realization $X(t)$ of the random process would have the formula

$$X(t) = A\cos(\Omega t + \Phi),\tag{4.44}$$

with A, Ω, and Φ chosen independently at random. Figure 4.1 generically displays every possible realization.

The random sine wave (as it is called, even though we are using a cosine) certainly doesn't look very noisy. Indeed, prediction is almost trivial since the curve can be

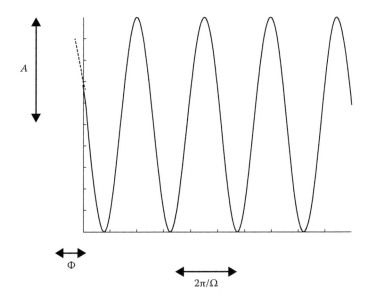

FIGURE 4.1 Random sine wave $X(t)$ versus t.

---

* Note, however, that the first three equations in (4.42) are linear and can be solved for the autoregressive coefficients $a(k)$ easily.

completely determined by four suitable measurements (Problem 8). It doesn't look stationary either. But it serves as a fairly simple academic example for the evaluation of means and autocorrelations, since the calculation of the expected values reduces mainly to trigonometric identities.

Let us assume that the pdf* of the amplitude A is given by $f_A(a)$ for $a > 0$, the pdf for Ω is $f_\Omega(\omega)$ for $\omega > 0$, and for Φ it is $f_\Phi(\varphi)$ for $0 < \varphi < 2\pi$. Then the (time-dependent) mean and autocorrelation are given by

$$\mu_{\text{sine}}(t) = E\{X(t)\} = \int_0^\infty \int_0^\infty \int_0^{2\pi} a\cos(\omega t + \varphi) f_A(a) f_\Omega(\omega) f_\Phi(\varphi)\, d\varphi\, d\omega\, da$$

$$R_{\text{sine}}(t_1, t_2) = E\{X(t_1)X(t_2)\}$$

$$= \int_0^\infty \int_0^\infty \int_0^{2\pi} a^2 \cos(\omega t_1 + \varphi)\cos(\omega t_2 + \varphi) f_A(a) f_\Omega(\omega) f_\Phi(\varphi)\, d\varphi\, d\omega\, da$$

$$= \int_0^\infty \int_0^\infty \int_0^{2\pi} a^2 \frac{\cos\omega\big[t_1 - t_2\big] + \cos\big(\omega\big[t_1 + t_2\big] + 2\varphi\big)}{2} f_A(a) f_\Omega(\omega) f_\Phi(\varphi)\, d\varphi\, d\omega\, da.$$

These expressions become interesting when the phase Φ is "completely" random—that is, when its pdf $f_\Phi(\varphi)$ is uniform (equal to $1/2\pi$), because then the Φ integrations simplify to

$$\mu_{\text{sine}}(t) = E\{X(t)\} = \frac{1}{2\pi}\int_0^\infty \int_0^\infty a\, f_A(a) f_\Omega(\omega) \int_0^{2\pi} \cos(\omega t + \varphi)\, d\varphi\, d\omega\, da = 0,$$

$$R_{\text{sine}}(t_1,t_2) = \frac{1}{2\pi}\int_0^\infty a^2 f_A(a)\, da \int_0^\infty f_\Omega(\omega) \int_0^{2\pi} \frac{\cos\omega\big[t_1 - t_2\big] + \cos\big(\omega\big[t_1 + t_2\big] + 2\varphi\big)}{2}\, d\varphi\, d\omega$$

$$= \frac{1}{2\pi}\int_0^\infty a^2 f_A(a)\, da \int_0^\infty f_\Omega(\omega)\, 2\pi \frac{\cos\omega\big[t_1 - t_2\big] + (0)}{2}\, d\omega$$

$$= \frac{1}{2} E\{A^2\} \int_0^\infty f_\Omega(\omega)\cos\omega\big[t_1 - t_2\big]\, d\omega \tag{4.45}$$

and the process becomes zero-mean and (wide-sense) stationary (Section 3.1; R_{sine} depends on the difference $\tau_1 - \tau_2$.)! Thus by randomly shifting the starting point of the curve in Figure 4.1, we mask its pattern of peaks and zero crossings.

---

* *probability density function.*

Invoking the liberal interpretation of the Fourier transform for generalized functions, we find the PSD of the random sine wave (with uniformly distributed Φ) to be

$$S_{sine}(f) = \int_{-\infty}^{\infty} R_X(\tau)e^{-i2\pi f\tau}d\tau = \frac{1}{2}E\{A^2\}\int_0^{\infty} f_\Omega(\omega)\int_{-\infty}^{\infty}\cos\omega\tau\, e^{-i2\pi f\tau}\,d\tau\,d\omega$$

$$= \frac{1}{2}E\{A^2\}\int_0^{\infty} f_\Omega(\omega)\frac{\delta\left(\frac{\omega}{2\pi}-f\right)+\delta\left(\frac{\omega}{2\pi}+f\right)}{2}d\omega$$

$$= \frac{\pi}{2}E\{A^2\}f_\Omega(|2\pi f|). \tag{4.46}$$

In other words, $S_X(f)$ is proportional to the probability that the generator's frequency dial (Ω) was set to the frequency $|2\pi f|$.[*]

Perhaps more significantly, (4.46) demonstrates that *given any power spectral density $S_X(f)$, there is a random sine wave $A\cos(\Omega t+\Phi)$ having this PSD*; you can show that this is accomplished with the choice

$$A = \sqrt{4\int_0^{\infty} S_X(f)df}\text{ (which is }not\text{ random)}, \quad f_\Omega(\omega) = (2/\pi A^2)S_X(\omega/2\pi), \quad \omega > 0, \quad \text{and}$$

$f_\Phi(\varphi) = (1/2\pi), \quad 0 < \varphi < 2\pi.$

However, the random sine wave is decidedly not ergodic, because the time average of $X(t)X(t-\tau)$ equals

$$\lim_{T\to\infty}\frac{1}{2T}\int_{-T}^{T} X(t)X(t-\tau)dt = A^2\lim_{T\to\infty}\frac{1}{2T}\int_{-T}^{T}\cos(\Omega t+\Phi)\cos(\Omega[t-\tau]+\Phi)dt$$

$$= A^2\lim_{T\to\infty}\frac{1}{2T}\int_{-T}^{T}\frac{\cos\Omega\tau+\cos(\Omega[2t-\tau]+2\Phi)}{2}dt$$

$$= \begin{cases} A^2\dfrac{\cos\Omega\tau}{2} & \text{if } \Omega \neq 0, \\[2mm] A^2\dfrac{1+\cos 2\Phi}{2} & \text{if } \Omega = 0, \end{cases}$$

and it bears only a passing resemblance to $R_X(t_1, t_2)$ in (4.45). Therefore, the random sine wave may be unsuitable as a model.

ONLINE SOURCES

The properties of a random sine wave buried in noise is a classical problem in detection theory. One of the original sources is

Rice, S.O. Statistical properties of a sine wave plus random noise. Accessed June 24, 2016. http://ieeexplore.ieee.org/xpls/abs_all.jsp?arnumber=6767541&tag=1.

---

[*] Why absolute value? Bear in mind that Ω is the (angular) frequency of the signal generator and is taken to be positive, but f is the (cycle) frequency in the Fourier transform of R_X and runs from $-\infty$ to ∞.

4.8 THE BERNOULLI AND BINOMIAL PROCESSES

The simplest of all honestly* random processes is the *Bernoulli sequence*, described in Section 2.1. It consists of independent repetitions of an experiment with two outcomes—like a run of coin flips. It is clearly stationary (and ergodic). We assign the numerical values a and b to the two outcomes, with the probability of the outcome "a" being p and (of course) $1 - p$ is the probability of "b." The means and autocorrelations are trivial and the following table is easily verified:

SUMMARY: BERNOULLI PROCESS

$$\text{For } n = 1, 2, 3, \ldots, X(n) = \begin{cases} a \text{ with probability } p \\ b \text{ with probability } 1-p \end{cases}.$$

$$\mu_{Bernoulli} \equiv E\{X(n)\} = pa + (1-p)b$$

$$R_{Bernoulli}(n_1, n_2) \equiv E\{X(n_1)X(n_2)\} = \begin{cases} pa^2 + (1-p)b^2 & if \quad n_1 = n_2 \\ \mu_{Bernoulli}^2 & if \quad n_1 \neq n_2 \end{cases}$$

$$R_{Bernoulli}(m) = \mu_{Bernoulli}^2 + \left[pa^2 + (1-p)b^2 - \mu_{Bernoulli}^2 \right]\delta_{m0}$$

$$S_{Bernoulli}(f) = \mu_{Bernoulli}^2 \delta(f) + pa^2 + (1-p)b^2 - \mu_{Bernoulli}^2.$$

If we add up all the a's and b's that have occurred by the nth trial of a Bernoulli process, the totals comprise a *binomial process*.[†] The probability that after n trials we have a total of r a's and $(n - r)$ b's equals the probability of getting a's on a *specific set* of r trials, $p^r(1 - p)^{n-r}$, times the number of sets of size r in a collection of n trials, $\binom{n}{r}$ (see Section 1.4):

$$\text{Prob}\left\{ Y(n) \equiv \sum_{j=1}^{n} X(n) = ra + (n-r)b \right\} = \binom{n}{r} p^r (1-p)^{n-r}.$$

---

* I would not say the sine (4.44) is "honestly" random since four measurements remove all the uncertainty.
† Binomial, B.P. 1654. *Traité du triangle arithmétique*. Paris, France: Deprez.

The binomial process is nonstationary, since the totals increase as we perform more trials. In fact,

$$\mu_{binomial}(n) = E\{X_{binomial}(n)\} = E\left\{\sum_{r=1}^{n} X_{Bernoulli}(r)\right\} = n\mu_{Bernoulli} = n\left[pa + (1-p)b\right].$$

$$R_{binomial}(n_1, n_2) = E\{(X_{Bernoulli}(1) + X_{Bernoulli}(2) + \cdots + X_{Bernoulli}(n_1))$$

$$\times (X_{Bernoulli}(1) + X_{Bernoulli}(2) + \cdots + X_{Bernoulli}(n_2))\}.$$

If $n_1 \le n_2$, after multiplication there are n_1 terms on the right with equal indices and $n_1 n_2 - n_1$ terms with unequal indices. From the Bernoulli formulas, then

$$R_{binomial}(n_1, n_2) = n_1\left[pa^2 + (1-p)b^2\right] + (n_1 n_2 - n_1)\mu_{Bernoulli}^2$$

$$= n_1 n_2 \mu_{Bernoulli}^2 + \left[pa^2 + (1-p)b^2 - \mu_{Bernoulli}^2\right]\min(n_1, n_2). \quad (4.47)$$

The Galton machine pictured in Figure 4.2 runs a binomial process for each ball. The probabilities of the ball being diverted to the left or right at each post are 1/2, so its total excursion when it gets to the collector is described by the binomial distribution, and inevitably the bin accumulations match this distribution. Typically the normal distribution (not the binomial) is painted on the machine as the "target" for the bin accumulations. *When $np(1-p) \gg 1$, the binomial distribution is well approximated by the normal with the same mean and standard deviation.* The machine in Figure 4.2 appears to have more than $n = 20$ rows of scatterers, so $np(1-p) \ge 20 \times 0.5 \times 0.5 = 5$.

ONLINE SOURCES

The following sites provide excellent visualizations that allow user input for the probability p, the number of trials n, and single-trial or batch display.

Bernoulli trials:
Siegrist, K. Bernoulli trials. Random (formerly Virtual Laboratories in Probability and Statistics). Accessed June 24, 2016. http://www.math.uah.edu/stat/bernoulli/.
Brown, A. Dice Roller & Penny Flipper. Wolfram Demonstrations Project. Accessed June 24, 2016. http://demonstrations.wolfram.com/DiceRollerPennyFlipper/.
Chris, B. Successes and failures in a run of Bernoulli trials. Wolfram Demonstrations Project. Accessed June 24, 2016. http://demonstrations.wolfram.com/Successes AndFailuresInARunOfBernoulliTrials/.

---

* This was discovered by De Moivre. See Daw, R.H. and Pearson, E.S. 1972. *Studies in the History of Probability and Statistics.* Abraham De Moivre's 1733 derivation of the normal curve: A bibliographical note. *Biometrika* 59(3): 677–680.

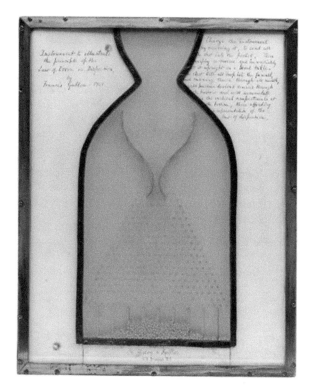

FIGURE 4.2 Galton machine. (Copyright UCL Galton Collection, University College London, London, U.K.)

Binomial process:
Savory, P. Binomial probability distribution. Wolfram Demonstrations Project. Accessed June 24, 2016. http://demonstrations.wolfram.com/BinomialProbabilityDistribution/.

Simulators for the Galton machine:
Vuilleumier, B. Idealized Galton Board. Wolfram Demonstrations Project. Accessed June 24, 2016. http://demonstrations.wolfram.com/IdealizedGaltonBoard/.
Schreiber, M. Flexible Galton Board. Wolfram Demonstrations Project. Accessed June 24, 2016. http://demonstrations.wolfram.com/FlexibleGaltonBoard/.

Demonstrations of the normal approximation to the binomial:
Boucher, C. Normal approximation to a binomial random variable. Wolfram Demonstrations Project. Accessed June 24, 2016. http://demonstrations.wolfram.com/NormalApproximationToABinomialRandomVariable/.

Published case studies:
Rajan, V. Bernoulli trials—Binomial experiment's usage in Fraud Detection. Data Science Central. Accessed June 24, 2016. http://www.datasciencecentral.com/profiles/blogs/bernoulli-trails-binomial-experiment-s-usage-in-fraud-detection.

(describes the use of Bernoulli trials to test hypotheses in the detection of the root causes of an *E. coli* outbreak [spinach] and of apparent taxicab frauds [unapproved booking apps])

Liu, S., Ding, W., Cohen, J.P., Simovici, D., and Stepinski, T. Bernoulli trials based feature selection for crater detection. ACM Digital Library. Accessed June 24, 2016. http://dl.acm.org/citation.cfm?id=2094048&dl=ACM&coll=DL&CFID=623181405&CFTOKEN=41569631.

(discusses the use of Bernoulli trials to enable automatic feature detection in a Martian terrain explorer)

Tutorial 5: Clinical trial. Microsoft Research. Accessed June 24, 2016. http://research.microsoft.com/en-us/um/cambridge/projects/infernet/docs/clinical%20trial%20tutorial.aspx.

(explores how Bernoulli reasoning is used in establishing the efficacy of drugs versus placebos in clinical trials)

Villeneuve, P.J. Binomial distribution. *Encyclopedia of Public Health*. Accessed June 24, 2016. http://www.encyclopedia.com/topic/Binomial_distribution.aspx.

(describes how the binomial process is used to analyze clinical trials in general)

SUMMARY: BINOMIAL PROCESS

$$X_{binomial}(n) = \sum_{k=1}^{n} X_{Bernoulli}(k)$$

$$\mu_{binomial}(n) = n\left[pa + (1-p)b\right].$$

$$R_{binomial}(n_1, n_2) = n_1 n_2 \mu_{Bernoulli}^2 + \left[pa^2 + (1-p)b^2 - \mu_{Bernoulli}^2\right]\min(n_1, n_2).$$

4.9 SHOT NOISE AND THE POISSON PROCESS

Most kinds of flow—fluid flow, current flow, etc.—are typically viewed as *continua*, in the sense that the fluid or charge is modeled by a continuous mass or charge density. So a uniform flow looks the same wherever/whenever it is viewed. At the atomic level, however, the flow is actually a conglomerate of discrete particles; and a true, detailed description of the flow would recognize a fluid motion as a drifting of a dense population of molecules rattling off of the container walls and each other, and a diode current as a rapid sputtering of individual electrons crossing the semiconductor junction. So there are minuscule random fluctuations even in a "uniform" flow, simply due to the discrete nature of the carriers.

Now the number of electrons crossing a diode junction per second in a 1 mA current is on the order of 10^{15} so that the average time lapse between emissions is 10^{-15} seconds. So, we can't observe the particulate nature of the flow. But it is basically a high-speed version of the same phenomenon as, say, customers arriving at a drive-in restaurant, or phone calls received by a business office. We attribute an average arrival rate to such phenomena—a "long-time" average—and regard the fluctuations as noise in the model. Such noise is known as *shot noise* (recall Example 2.4).

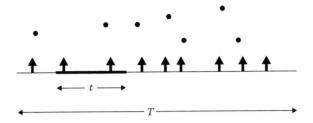

FIGURE 4.3 $m = 9$ events in an interval T.

To analyze this model, consider the statistical description of m events occurring randomly in T seconds—analogous to a handful of flower seeds scattered over a furrowed row of length T in a garden. See Figure 4.3.

The probability of a particular event landing in any interval I_t of length t is t/T. So, reasoning as we did for the binomial process, we deduce that the probability of n of the m events landing in the interval I_t is given by

$$\binom{m}{n}\left(\frac{t}{T}\right)^n\left(\frac{T-t}{T}\right)^{m-n}. \tag{4.48}$$

Now to model shot noise, let T and m go to infinity in this model while $m/T \equiv \lambda$, the long-time average number of events per second, stays constant. We process (4.48) judiciously, hoarding factors of m/T and isolating the remaining pieces so as to expose their limits:

Poisson probability: The probability of n events in an interval of length t is
the limit of

$$\frac{m!}{n!(m-n)!}\left(\frac{t}{T}\right)^n\left(\frac{T-t}{T}\right)^{m-n}$$

$$= \frac{t^n}{n!}\frac{m(m-1)(m-2)\cdots(m-n+1)}{T^n}\left(1-\frac{t}{T}\right)^{-n}\left(1-\frac{t}{T}\right)^m$$

$$= \frac{t^n}{n!}\left(\frac{m}{T}\right)^n\underbrace{\frac{m(m-1)(m-2)\cdots(m-n+1)}{m\ \ m\ \ m\ \ \cdots\ \ m}}\left(1-\frac{t}{T}\right)^{-n}\left(1-\frac{t}{T}\right)^{(T/t)(m/T)t}$$

$$\parallel \qquad\qquad \downarrow_{m\to\infty} \qquad\qquad \downarrow_{T\to\infty} \quad \downarrow_{T\to\infty}$$

$$\frac{(\lambda t)^n}{n!} \qquad\qquad 1 \qquad\qquad 1 \qquad e^{-\lambda t}$$

$$P(n;t) = \frac{(\lambda t)^n}{n!}e^{-\lambda t}. \tag{4.49}$$

As a check, note that

$$\sum_{n=0}^{\infty} \frac{(\lambda t)^n}{n!} e^{-\lambda t} = 1 \quad \left(\text{since} \sum_{n=0}^{\infty} \frac{(\lambda t)^n}{n!} = e^{\lambda t} \right). \qquad (4.50)$$

So the **Poisson process*** $X_{Poisson}(t)$, the *accumulated* number of shot noise events in the interval $\{0, t\}$, is a discrete-valued process defined on the time continuum with a pdf given by

$$f_{X_{Poisson}(t)}(x) = \sum_{n=0}^{\infty} \frac{(\lambda t)^n}{n!} e^{-\lambda t} \delta(x-n).$$

In Figure 4.4, we have depicted the individual events as delta functions, so $X_{Poisson}(t)$ equals the integral, from 0 to t, of the shot noise.

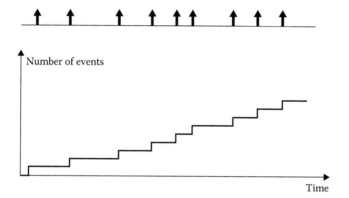

FIGURE 4.4   The Poisson process.

According to our interpretation of the longtime average, the expected number of events in t seconds should be λt; and this is confirmed by the calculation

$$\mu_{Poisson}(t) = E\{X_{Poisson}(t)\} = \int_{-\infty}^{\infty} x f_{X_{Poisson}(t)}(x) dx$$

$$= \sum_{n=0}^{\infty} n \frac{(\lambda t)^n}{n!} e^{-\lambda t} = \lambda t \sum_{n=1}^{\infty} \frac{(\lambda t)^{n-1}}{(n-1)!} e^{-\lambda t} = \lambda t \left\{ \sum_{r=0}^{\infty} \frac{(\lambda t)^r}{r!} e^{-\lambda t} \right\} = \lambda t \cdot 1 \qquad (4.51)$$

---

* Poisson, S.D. 1837. *Probabilité des jugements en matière criminelle et en matière civile, précédées des règles générales du calcul des probabilités.* Paris, France: Bachelier.
  Sometimes called the Poisson counting process, and sometimes called the law of rare events because of the particulate nature depicted in Figure 4.2. Note that (4.49) is not a pdf.

There is an additional wrinkle in the Poisson process. If $t = T/2$, according to (4.48) for $m = 5$ "seeds," the probability of 3 falling in the left "half-garden" and 2 falling in the right is $10/2^5$, whereas the probability of three seeds falling in the left half-garden, *given that there are* two *seeds in the right half-garden*, is one; the events are not independent. However, in the limit the number of events in each disjoint interval becomes independent: for example, with some juggling, we find

probability of 3 events in $\left[0, t\right]$ and 2 events in $\left[t, 2t\right]$

$$= \lim_{m \to \infty, \, T \to \infty, \, (m/T) = \lambda} \frac{m(m-1)(m-2)(m-3)(m-4)}{3!2!} \left(\frac{t}{T}\right)^3 \left(\frac{t}{T}\right)^2 \left(\frac{T-2t}{T}\right)^{m-5}$$

$$= \lim \frac{m(m-1)(m-2)(m-3)(m-4)}{m^5} \frac{(mt/T)^5}{3!2!} \left(1-\frac{2t}{T}\right)^{-5} \left(1-\frac{2t}{T}\right)^{(T/2t)(2tm/T)}$$

$$\to (1) \frac{(\lambda t)^5}{3!2!} e^{-2t\lambda} = \left[\frac{(\lambda t)^3}{3!} e^{-t\lambda}\right]\left[\frac{(\lambda t)^2}{2!} e^{-t\lambda}\right]$$

$$= \{\text{probability of 3 events in } \left[0, t\right]\}\{\text{probability of 2 events in } \left[t, 2t\right]\}.$$

We say that the Poisson process has **independent increments**.

Further insight into shot noise and the Poisson processes is gained by posing the *waiting time problem*: What is the probability that the first event occurs after waiting t seconds? Of course, since t is a continuous variable, the sensible question is actually the following:

What is the probability density function $f(t)$ for the time of arrival of the first event—so that $f(t)\, dt$ is the probability that the first event occurs between times t and $t + dt$?

We answer this by noting that the probability of one event in the interval $\{t, t + dt\}$ equals, by (4.49), $(\lambda dt)^1 e^{-\lambda dt}/1!$. And this will be the *first* event if there were no events in $\{0, t\}$. Therefore, since the intervals are disjoint, by the customary infinitesimal reasoning,*

$$f(t)\, dt = \frac{\lambda\, dt\, e^{-\lambda dt}}{1!} \frac{(\lambda t)^0 e^{-\lambda t}}{0!} = \lambda e^{-\lambda t} dt\, e^{-\lambda dt} = \lambda e^{-\lambda t} dt + O\left(dt^2\right).$$

Waiting time problem: The probability density function for the time until the first occurrence of an event is the **exponential distribution** $f(t) = \lambda e^{-\lambda t}$.

Similarly, the pdf for the time of arrival of the nth event is derived by noting, as before, that the probability of one event in the interval $\{t, t + dt\}$ equals, by (4.49), $(\lambda dt)^1 e^{-\lambda dt}/1!$, and this will be the nth event if there were $n-1$ events in $\{0, t\}$. Therefore,

$$f(t)\, dt = \frac{\lambda\, dt\, e^{-\lambda dt}}{1!} \frac{(\lambda t)^{n-1} e^{-\lambda t}}{(n-1)!}.$$

---

* $O(v)$ is the notation for any function that approaches zero, as $v \to 0$, so fast that $O(v)/v$ is bounded.

The pdf for the time until the nth event is the **Erlang distribution**: $f_{Erlang}(t) = \lambda(\lambda t)^{n-1}e^{-\lambda t}/(n-1)!$.

Interestingly, one can prove the converse: *If* the waiting time between events is exponentially distributed, then the process is Poisson; see Problem 22.

Having found the mean of the Poisson process (4.51), we turn to its autocorrelation:

$$R_{Poisson}(t_1, t_2) \equiv E\{X(t_1)X(t_2)\} = \sum_{n_1,n_2} n_1 n_2 \, \mathrm{Prob}\{X(t_1) = n_1 \ \& \ X(t_2) = n_2\}.$$

If the times are ordered as $0 < t_1 < t_2$, then we interpret this scenario as n_1 events occurring in $\{0, t_1\}$ and $0 \le n_2 - n_1 \equiv p$ events occurring in $\{t_1, t_2\}$. The independent increment property guarantees

$$R_{Poisson}(t_1, t_2) = \sum_{n_1=0}^{\infty}\sum_{p=0}^{\infty} n_1(n_1+p)\frac{(\lambda t_1)^{n_1}}{n_1!}e^{-\lambda t_1}\frac{(\lambda[t_2-t_1])^{p}}{p!}e^{-\lambda[t_2-t_1]},$$

and more clever algebra reveals that this equals

$R_{Poisson}(t_1, t_2)$

$$= \sum_{n_1=0}^{\infty} n_1^2 \frac{(\lambda t_1)^{n_1}}{n_1!}e^{-\lambda t_1} \sum_{p=0}^{\infty} \frac{(\lambda[t_2-t_1])^{p}}{p!}e^{-\lambda[t_2-t_1]}$$

$$+ \sum_{n_1=0}^{\infty} n_1 \frac{(\lambda t_1)^{n_1}}{n_1!}e^{-\lambda t_1} \sum_{p=0}^{\infty} p\frac{(\lambda[t_2-t_1])^{p}}{p!}e^{-\lambda[t_2-t_1]}$$

$$= \sum_{n_1=0}^{\infty} n_1^2 \frac{(\lambda t_1)^{n_1}}{n_1!}e^{-\lambda t_1}(1) + \sum_{n_1=0}^{\infty} n_1 \frac{(\lambda t_1)^{n_1}}{n_1!}e^{-\lambda t_1}\left(\lambda[t_2-t_1]\right) \quad \left(\text{as in}\,(4.50)\,\text{and}\,(4.51)\right)$$

$$= \sum_{n_1=0}^{\infty} n_1^2 \frac{(\lambda t_1)^{n_1}}{n_1!}e^{-\lambda t_1} + \lambda t_1 \left(\lambda[t_2-t_1]\right) \quad \left(\text{as in}\,(4.51)\right)$$

$$= \sum_{n_1=0}^{\infty} n_1 \frac{(\lambda t_1)^{n_1}}{(n_1-1)!}e^{-\lambda t_1} + \lambda t_1\left(\lambda[t_2-t_1]\right)$$

$$= \lambda t_1 \sum_{n_1=0}^{\infty} (n_1-1)\frac{(\lambda t_1)^{n_1-1}}{(n_1-1)!}e^{-\lambda t_1} + \lambda t_1 \sum_{n_1=0}^{\infty} (1)\frac{(\lambda t_1)^{n_1-1}}{(n_1-1)!}e^{-\lambda t_1} + \lambda t_1\left(\lambda[t_2-t_1]\right)$$

$$= (\lambda t_1)(\lambda t_1) + \lambda t_1(1) + \lambda t_1\left(\lambda[t_2-t_1]\right) = \lambda t_1 + \lambda^2 t_1 t_2 \quad \left(\text{by}\,(4.51)\,\text{and}\,(4.50)\right).$$

If $t_2 \le t_1$, then the subscripts in the formula become reversed, and we have

$$R_{Poisson}(t_1, t_2) = \lambda \min(t_1, t_2) + \lambda^2 t_1 t_2. \tag{4.52}$$

The autocorrelation is

$$C_{Poisson}\left(t_1, t_2\right) = R_{Poisson}\left(t_1, t_2\right) - E\left\{X\left(t_1\right)\right\}E\left\{\left(X\left(t_2\right)\right)\right\}$$

$$= \lambda \min\left(t_1, t_2\right) + \lambda^2 t_1 t_2 - \left(\lambda t_1\right)\left(\lambda t_2\right) = \lambda \min\left(t_1, t_2\right).$$

Thus, the variance of $X(t)$ is λt—the same as the mean.

Example 4.1: The Random Telegraph Signal

If $X(t) = \pm a$ changes sign each time a Poisson event occurs, the resulting process is known as the *random telegraph signal*. Clearly its mean is zero. To calculate its autocorrelation, note that $X(t_1)\,X(t_2) = a^2$ if there are an even number of events between $t = t_1$ and $t = t_2$, and otherwise $X(t_1)\,X(t_2) = -a^2$. Thus, for $t_2 = t_1 + \tau$ ($\tau \geq 0$),

$$R_{telegraph}\left(t_1, t_2\right) = E\left\{X\left(t_1\right)X\left(t_1+\tau\right)\right\} = +a^2 \sum_{n=0,2,\ldots} \frac{\left(\lambda\tau\right)^n}{n!} e^{-\lambda\tau} - a^2 \sum_{n=1,3,\ldots} \frac{\left(\lambda\tau\right)^n}{n!} e^{-\lambda\tau}$$

$$= a^2 \sum_{n=0}^{\infty} \frac{\left(-\lambda\tau\right)^n}{n!} e^{-\lambda\tau} = a^2 e^{-2\lambda\tau} \sum_{n=0}^{\infty} \frac{\left(-\lambda\tau\right)^n}{n!} e^{+\lambda\tau} = a^2 e^{-2\lambda\tau}$$

(recall (4.50)). The random telegraph signal is (second-order) stationary and its PSD[*] is the **Cauchy–Lorentz density:**[†]

$$S_{telegraph}\left(f\right) = a^2 \frac{\lambda}{\left(\pi f\right)^2 + \lambda^2}.$$

The random telegraph signal process occurs in semiconductor components where we see jumps in voltage levels occurring randomly (recall Figure 2.5). This is attributed to impurity-induced generation–recombination centers that trap carriers and hold them for exponential waiting times. The process, which may involve more than two levels, is known as "burst noise" or "popcorn noise" (imagine the voltage in Figure 2.5 driving a speaker).

(Another nuisance effect observed in semiconductor devices is known as $1/f$ noise because of the shape of its PSD graph. Noting the mathematical identity

$$\int_{-\infty}^{\infty} S_{telegraph}\left(f\right)\left(\frac{1}{\lambda}\right)d\lambda = \int_{-\infty}^{\infty} a^2 \frac{\lambda}{\left(\pi f\right)^2 + \lambda^2} \frac{d\lambda}{\lambda} = \frac{a^2}{f},$$

McWorter[‡] attributed $1/f$ noise to a superposition of traps on the device surface having mean waiting times $1/\lambda$, distributed with a density proportional to $1/\lambda$.

---

[*] *power spectral density.*

[†] Possibly first studied by Poisson: see Stigler, S.M. (1999). Cauchy and the Witch of Agnesi. In *Statistics on the Table*, Chapter 18. Cambridge, MA: Harvard.

[‡] McWhorter, A.L. 1957. 1/f noise and germanium surface prosperities. In *Semiconductor Surface Physics*, ed. Kingston, R.H., pp. 207–228. Philadelphia, PA: University of Pennsylvania Press.

Hooge,* on the other hand, attributed 1/*f* noise to carrier scattering on the crystal lattice. Other hybrid phenomenological characterizations have been proposed, but no simple physical model for 1/*f* noise has achieved widespread acceptance.)

As is obvious from Figure 4.3 as well as (4.51) and (4.52), the Poisson process is not stationary (so we don't seek its PSD); it is the accumulation of shot noise, and it increases by 1 each time a shot noise event occurs. What about the shot noise itself—the individual events? Can we even consider it as a random process—as a series of delta functions, occurring at random times? Can we attribute a mean value to it? Certainly not by the usual definition; what sense could we make of

$$E\{X_{shot}(t)\} = 0 \times \{\text{Probability that } X_{shot}(t) = 0\} + \infty \times \{\text{Probability that } X_{shot}(t) = \infty\}?$$

However, since we regarded the Poisson process as an integral of the shot noise (Figure 4.4), we shall formally regard shot noise as the derivative of the Poisson process:

$$X_{shot}(t) = \frac{d}{dt} X_{Poisson}(t). \tag{4.53}$$

Now, if a random process $X(t)$ is truly differentiable, then

$$\frac{dX(t)}{dt} = \lim_{\Delta t \to 0} \frac{X(t+\Delta t) - X(t)}{\Delta t}, \tag{4.54}$$

and we might reasonably expect the means to be simply related:

$$E\left\{\frac{dX(t)}{dt}\right\} = E\left\{\lim_{\Delta t \to 0} \frac{X(t+\Delta t) - X(t)}{\Delta t}\right\}$$

$$= \lim_{\Delta t \to 0} E\left\{\frac{X(t+\Delta t) - X(t)}{\Delta t}\right\}$$

$$= \lim_{\Delta t \to 0} \left\{\frac{E\{X(t+\Delta t)\} - E\{X(t)\}}{\Delta t}\right\} = \frac{d}{dt} E\{X(t)\}. \tag{4.55}$$

If we apply (4.55) to (4.53), we find that we can assign a "mean" to the shot noise process:

$$\mu_{shot} = E\{X_{shot}(t)\} = E\left\{\frac{d}{dt} X_{Poisson}(t)\right\} = \frac{d}{dt} E\{X_{Poisson}(t)\} = \frac{d}{dt}(\lambda t) = \lambda, \tag{4.56}$$

the value we originally ascribed to the average number of arrivals per unit time.

---

* Hooge, F.N. 1976. 1/f noise. *Physica* 83B(1):14.

Accepting the validity of (4.55), we reason similarly to calculate the autocorrelation of the derivative. The key to the argument is to note the simple identity

$$\frac{dX(t_1)}{dt_1} \frac{dX(t_2)}{dt_2} = \frac{\partial}{\partial t_1} \left[\frac{\partial}{\partial t_2} X(t_1) X(t_2) \right], \tag{4.57}$$

for then we can propose that

$$R_{dX/dt}(t_1, t_2) \equiv E\left\{ \frac{dX(t_1)}{dt_1} \frac{dX(t_2)}{dt_2} \right\} = E\left\{ \frac{\partial}{\partial t_1} \left[\frac{\partial}{\partial t_2} X(t_1) X(t_2) \right] \right\}$$

$$= \frac{\partial}{\partial t_1} \left[\frac{\partial}{\partial t_2} E\{ X(t_1) X(t_2) \} \right]$$

$$= \frac{\partial}{\partial t_1} \frac{\partial}{\partial t_2} R_X(t_1, t_2). \tag{4.58}$$

To calculate the autocorrelation according to (4.58), we will need to take mixed derivatives of

$$R_{Poisson}(t_1, t_2) = \begin{cases} \lambda t_1 + \lambda^2 t_1 t_2 & \text{if } t_2 \geq t_1, \\ \lambda t_2 + \lambda^2 t_1 t_2 & \text{if } t_2 \leq t_1. \end{cases} \tag{4.59}$$

Thus, we have

$$\frac{\partial R_{Poisson}(t_1, t_2)}{\partial t_2} = \begin{cases} \lambda^2 t_1 & \text{if } t_2 \geq t_1 \\ \lambda + \lambda^2 t_1 & \text{if } t_2 \leq t_1 \end{cases} = \lambda u(t_1 - t_2) + \lambda^2 t_1$$

where $u(t)$ is the unit step function.* Therefore, by (4.57),

$$R_{shot}(t_1, t_2) = \frac{\partial}{\partial t_1} \frac{\partial R_{Poisson}(t_1, t_2)}{\partial t_2} = \lambda \delta(t_1 - t_2) + \lambda^2 = R_{shot}(|t_1 - t_2|). \tag{4.60}$$

Thus, the shot noise process (with our liberal interpretation) has constant mean (4.56) and autocorrelation depending only on the (magnitude of the) time difference τ—it is second-order stationary. Its power spectrum is found by taking the Fourier transform of (4.60):

$$S_{shot}(f) = \int_{-\infty}^{\infty} R_{shot}(\tau) e^{-j2\pi f \tau} d\tau = \lambda^2 \delta(f) + \lambda. \tag{4.61}$$

---

* $u(t) = 1$ if $t > 0$, 0 otherwise.

The first term, contributing only for $f = 0$, represents the power in the average, or DC, component of the flow. The second represents white noise. (Recall Section 3.6.)

Example 4.2: Current Noise

If we interpret the shot events as electron emissions from a cathode and each electron carries a charge of q Coulombs, then the total charge emitted in t seconds equals $qX_{Poisson}(t)$. Its derivative, $qX_{shot}(t)$, equals the instantaneous current $I(t)$. It has mean

$$E\{I(t)\} = E\left\{q\frac{d}{dt}X_{Poisson}(t)\right\} = q\lambda = I_{DC},$$

the "DC" current, and autocorrelation

$$R_I(t_1,t_2) \equiv E\left\{\frac{d[qX_{Poisson}(t_1)]}{dt_1}\frac{d[qX_{Poisson}(t_2)]}{dt_2}\right\} = q^2\lambda\delta(\tau) + q^2\lambda^2 = qI_{DC}\delta(\tau) + I_{DC}^2.$$

(4.62)

The power spectrum density of the current, $S_I(f) = I_{DC}^2\delta(f) + qI_{DC}$, thus contains the familiar DC component at 0 Hz and white noise of intensity qI_{DC} distributed uniformly in frequency. It is customary for electrical engineers to combine the negative and positive frequency components, quoting a (white) shot noise power of $2qI_{DC}$ Watts per Hertz.

ONLINE SOURCES AND DEMONSTRATIONS

Poisson process:
Ruskeepää, H. Simulating the Poisson process. Wolfram Demonstrations Project. Accessed June 24, 2016. http://demonstrations.wolfram.com/SimulatingThePoissonProcess/.
Cicilio, J. Two-state random walk distribution. Wolfram Demonstrations Project. Accessed June 24, 2016. http://demonstrations.wolfram.com/TwoStateRandomWalkDistribution/.
The Poisson Distribution. University of Massachusetts at Amherst. Accessed June 24, 2016. https://www.umass.edu/wsp/resources/poisson/.
    [describes one of the original applications of Poisson processes, predicting the number of Prussian cavalrymen kicked to death by their horses]

4.10   RANDOM WALKS AND THE WIENER PROCESS

Consider a binomial process wherein a man flips a coin every τ seconds (independent Bernoulli trials); if the flip comes up heads, he takes a step of length s to the right, and otherwise he takes the same-sized step to the left. This process is known as a **random walk**. Figure 4.5 compares the random walk with the Poisson process.

Applying the equations of Section 4.8 with $a = s$, $b = -s$, we find

$$\mu_{RandWalk}(n\tau) = n(2p-1)s,$$

(4.63)

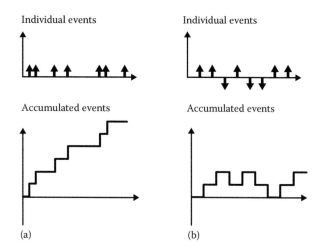

FIGURE 4.5 (a) Poisson process and (b) random walk.

$$R_{RandWalk}\left(n_1\tau,\ n_2\tau\right) = n_1 n_2 \left(2p-1\right)^2 s^2 + 4p\left(1-p\right)s^2 \min\left(n_1,n_2\right). \qquad (4.64)$$

Note that for a "fair" coin, $p = 1/2$ and these formulas simplify to

$$\mu_{RandWalk}\left(n\tau\right) = 0, \quad R_{RandWalk}\left(n_1\tau,n_2\tau\right) = s^2 \min\left(n_1,n_2\right). \qquad (4.65)$$

The **Wiener process*** is a continuous version of the random walk process with $p = 1/2$. We will let s and τ approach be zero so that $\{n\tau\}$ and x become continua. In anticipation of this we associate, with each real time t, that multiple of τ that barely precedes (or equals) it,[†] $n\tau = \lfloor t/\tau \rfloor \tau$. Then formally, we have (for $p = 1/2$)

$$X_{Wiener}\left(t\right) = \lim_{\tau \to 0} X_{RandWalk}\left(n\tau = \lfloor t/\tau \rfloor \tau\right),$$

$$E\left\{X_{Wiener}\left(t\right)\right\} = \lim_{\tau \to 0} E\left\{X_{RandWalk}\left(n\tau = \lfloor t/\tau \rfloor \tau\right)\right\} = 0,$$

$$R_{Wiener}\left(t_1,t_2\right) = \lim_{s,\tau \to 0} R_{RandWalk}\left(\lfloor t_1/\tau \rfloor \tau, \lfloor t_2/\tau \rfloor \tau\right) = \lim_{s,\tau \to 0} s^2 \min\left(t_1/\tau, t_2/\tau\right)$$

$$= \lim_{s,\tau \to 0} \frac{s^2}{\tau} \min\left(t_1,\ t_2\right)$$

$$= \alpha \min\left(t_1,\ t_2\right)$$

---

* Wiener, N. 1921. The average of an analytical functional and the Brownian movement. *Proc. Natl. Acad. Sci. U.S.A.* 7: 294–298.
[†] $\lfloor x \rfloor$ is the highest integer $\leq x$.

if we maintain the ratio s^2/τ equal to α as we take the limit. In fact, since $X_{Wiener}(t)$ is the limit of a sum of independent identically distributed random variables, the Central Limit Theorem can be used to argue that its pdf is Gaussian $N(0, \sqrt{\alpha t})$:

$$f_{X_{Wiener}(t)}(x) = \frac{e^{-\frac{(x-\mu)^2}{2\sigma^2}}}{\sqrt{2\pi\sigma^2}} = \frac{e^{-\frac{x^2}{2\alpha t}}}{\sqrt{2\pi\alpha t}}.$$

This is precisely the Green's function for the diffusion equation, with α equal to 2 times the diffusivity. The analogy between random walk and heat propagation has been exploited in studies of both areas.*

The Wiener process serves as a model for **Brownian motion**, the random fluctuations produced in the motion of a tiny particle suspended in a fluid, by molecular collisions. Although its mean is zero, the Wiener process is not stationary; its variance $R_{Wiener}(t, t) = \alpha t$ grows with time. However, its (formal) derivative is stationary, having mean and autocorrelation given by

$$E\left\{\frac{dW(t)}{dt}\right\} = 0, \; R_{dW/dt}(t_1, t_2) = \frac{\partial}{\partial t_1}\frac{\partial}{\partial t_2} R_{Wiener}(t_1, t_2) = \alpha\delta(t_1 - t_2)$$

(recall a similar calculation for shot noise). The power spectral density of dW/dt is constant (α), so it is another example of white noise. dW/dt is often referred to as **white Gaussian noise**, "WGN," and it is used to model Johnson noise (thermal-noise voltage; recall Example 2.3).

ONLINE SOURCES

The following sites provide excellent visualizations of the random walk process:

Ruskeepää, H. Simulating the simple random walk. Wolfram Demonstrations Project. Accessed June 24, 2016. http://demonstrations.wolfram.com/SimulatingTheSimpleRandomWalk.
Siegrist, K. Bernoulli trials. Random (formerly Virtual Laboratories in Probability and Statistics). Accessed June 24, 2016. http://www.math.uah.edu/stat/bernoulli/.

For a Wiener process simulator, set the correlation to zero at

Kozlowski, A. Correlated Wiener processes. Wolfram Demonstrations Project. Accessed June 24, 2016. http://demonstrations.wolfram.com/CorrelatedWienerProcesses/.

To see white Gaussian noise set $H = 0.5$ at

McLeod, I. Fractional Gaussian noise. Wolfram Demonstrations Project. Accessed June 24, 2016. http://demonstrations.wolfram.com/FractionalGaussianNoise/.

---

* Lawler, G.F. 2010. *Random Walk and the Heat Equation*. American Mathematical Society (Student Mathematical Library 55).

4.11 MARKOV PROCESSES

As we indicated in Section 2.2, the complete characterization of a random process entails knowledge of all the probability distribution functions:

$$f_{X(t)}(x), f_{X(t_2)|X(t_1)}(x_2|x_1), f_{X(t_3)|X(t_1)\&X(t_2)}(x_3|x_1,x_2),\ldots$$
$$f_{X(t_1),X(t_2)}(x_1,x_2), f_{X(t_1),X(t_2),X(t_3)}(x_1,x_2,x_3),\ldots.$$

Markov processes* are distinguished by the fact that the conditional pdf of $X(t)$, given values at earlier times $\{X(t_1), X(t_2), ..., X(t_k)\}$, depends solely on the most recent value. In other words, if $t_1 < t_2 < \cdots < t_k < t$ and one is given the values of X at these earlier times, the conditional probability distribution for $X(t)$ depends only on $X(t_k)$:

$$f_{X(t)|X(t_1)\&X(t_2)\&...\&X(t_k)}(x|x_1,x_2,...,x_k) = f_{X(t)|X(t_k)}(x|x_k).$$

The prior history, $X(t_1), X(t_2), ..., X(t_{k-1})$, can be discarded.

Poisson and Wiener processes are Markov; if $X(t_k)$ is known, the subsequent *increments* to X are modeled by the probability distribution functions we have derived and simply added to $X(t_k)$, while knowledge of previous values are irrelevant. The equation for the ARMA(1,0) process (recall Section 4.3),

$$X(n) = a(1)X(n-1) + b(0)V(n),$$

shows immediately that knowledge of X at any one particular time determines its pdf at later times, and additional knowledge of X at earlier times contributes no further information. Markov processes have been used to model birth–death and epidemic phenomena, and rental-car movement.

Many excellent studies have been dedicated to aspects of Markov processes. Herein we provide a sampling of the richness of the theory by considering *discrete stationary* processes where $X(n)$ can only take on a finite set of values. As an example, suppose you are watching a game where a marker moves around the corners of a square (Figure 4.5). At each second the probability of its moving clockwise from corner #1 is 0.3, the probability of moving counterclockwise is 0.2, and with probability 0.5 it remains on corner #1. The other transition probabilities are displayed in Figure 4.6, using loop-arrows to depict the stay-at-home choices.

Note that this game has the defining Markov property—given a history of values such as $X(1) = 3$, $X(2) = 2$, $X(5) = 4$, and $X(7) = 3$, the probabilities of future states depend only on the last datum, $X(7) = 3$. For example, the probabilities of the various states for $X(8)$, given that $X(7) = 3$, are

Prob$\{X(8) = 1 \mid X(7) = 3\} = 0$ (impossible),
Prob$\{X(8) = 2 \mid X(7) = 3\} = 0.2$ (counterclockwise),

---

* Markov, A.A. 1906. Rasprostranenie zakona bol'shih chisel na velichiny, zavisyaschie drug ot druga. *Izvestiya Fiziko-matematicheskogo obschestva pri Kazanskom universitete*, 2-ya seriya, tom 15: 135–156.

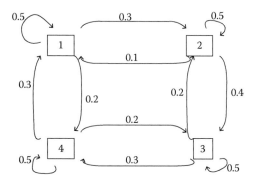

FIGURE 4.6 Markov process.

Prob$\{X(8) = 3 \mid X(7) = 3\} = 0.5$ (stay-at-home),
Prob$\{X(8) = 4 \mid X(7) = 3\} = 0.3$ (clockwise).

And the additional information about $X(1)$, $X(2)$, and $X(5)$ do not affect these probabilities. The process is *discrete* because it changes every second, rather than continuously; it is *stationary* because the same set of transition probabilities apply every turn.

From the given conditional probabilities, we can compute the *a priori* probabilities for $X(8)$ from those for $X(7)$ using the additive and conditional laws of probability (Sections 1.2 and 1.3). For example,

$$
\begin{aligned}
&\text{Prob}\{X(8) = 1\} \\
&= \text{Prob}\{X(8) = 1 \ \& \ X(7) = 1\} + \text{Prob}\{X(8) = 1 \ \& \ X(7) = 2\} \\
&\quad + \text{Prob}\{X(8) = 1 \ \& \ X(7) = 3\} + \text{Prob}\{X(8) = 1 \ \& \ X(7) = 4\} \\
&= \text{Prob}\{X(8) = 1 \mid X(7) = 1\} \times \text{Prob}\{X(7) = 1\} \\
&\quad + \text{Prob}\{X(8) = 1 \mid X(7) = 2\} \times \text{Prob}\{X(7) = 2\} \\
&\quad + \text{Prob}\{X(8) = 1 \mid X(7) = 3\} \times \text{Prob}\{X(7) = 3\} \\
&\quad + \text{Prob}\{X(8) = 1 \mid X(7) = 4\} \times \text{Prob}\{X(7) = 4\}. \quad\quad (4.66)
\end{aligned}
$$

Clearly matrix notation would be useful in encapsulating these relations. Write the *a priori* pdf for X at time n as a column vector $P(n)$:

$$
P(n) = \begin{bmatrix} \text{Prob}\{X(n) = 1\} \\ \text{Prob}\{X(n) = 2\} \\ \text{Prob}\{X(n) = 3\} \\ \text{Prob}\{X(n) = 4\} \end{bmatrix},
$$

and define the transition probability matrix $\mathbf{P}^{(1)}$ whose i, jth entry is the probability of jumping from corner j to corner i in one transition:

$$\mathbf{P}_{ij}^{(1)} = \text{Prob}\{X(n) = i \mid X(n-1) = j\},$$

$$\mathbf{P}^{(1)} = \begin{bmatrix} p_{11} & p_{12} & p_{13} & p_{14} \\ p_{21} & p_{22} & p_{23} & p_{24} \\ p_{31} & p_{32} & p_{33} & p_{34} \\ p_{41} & p_{42} & p_{43} & p_{44} \end{bmatrix} = \begin{bmatrix} 0.5 & 0.3 & 0 & 0.2 \\ 0.2 & 0.5 & 0.3 & 0 \\ 0 & 0.2 & 0.5 & 0.3 \\ 0.3 & 0 & 0.2 & 0.5 \end{bmatrix}. \tag{4.67}$$

Then (4.66) corresponds to the first row of the matrix equation,

$$\begin{bmatrix} \text{Prob}\{X(8) = 1\} \\ \text{Prob}\{X(8) = 2\} \\ \text{Prob}\{X(8) = 3\} \\ \text{Prob}\{X(8) = 4\} \end{bmatrix} \equiv P(8) = \begin{bmatrix} p_{11} & p_{12} & \cdots & p_{14} \\ p_{21} & p_{22} & \cdots & p_{24} \\ & & \ddots & \\ p_{41} & p_{42} & \cdots & p_{44} \end{bmatrix} \bullet P(7),$$

and since the transition probabilities do not change, the generalization is

$$P(n+1) = \mathbf{P}^{(1)} P(n). \tag{4.68}$$

Iteration of (4.68) shows that the two-transition probability matrix $\mathbf{P}^{(2)}$, giving the probability of going from corner i to corner j in *two* moves, is $\mathbf{P}^{(1)}$ squared:

$$P(n+2) = \mathbf{P}^{(1)} P(n+1) = \{\mathbf{P}^{(1)}\}^2 P(n) = \mathbf{P}^{(2)} P(n). \tag{4.69}$$

Thus, the two-transition probabilities for the four-corner game are

$$P^2 = \begin{bmatrix} 0.5 & 0.3 & 0 & 0.2 \\ 0.2 & 0.5 & 0.3 & 0 \\ 0 & 0.2 & 0.5 & 0.3 \\ 0.3 & 0 & 0.2 & 0.5 \end{bmatrix} \begin{bmatrix} 0.5 & 0.3 & 0 & 0.2 \\ 0.2 & 0.5 & 0.3 & 0 \\ 0 & 0.2 & 0.5 & 0.3 \\ 0.3 & 0 & 0.2 & 0.5 \end{bmatrix} = \begin{bmatrix} 0.37 & 0.30 & 0.13 & 0.20 \\ 0.20 & 0.37 & 0.30 & 0.13 \\ 0.13 & 0.20 & 0.37 & 0.30 \\ 0.30 & 0.13 & 0.20 & 0.37 \end{bmatrix}.$$

As a rather trivial consequence, we observe that for $p > n > m$

$$\text{Prob}\{X(p) = i \mid X(m) = j\} = \sum_k \text{Prob}\{X(p) = i \mid X(n) = k\} \text{Prob}\{X(n) = k \mid X(m) = j\},$$

$$\tag{4.70}$$

summed over all states $\{k\}$. In fact, you should see that (4.70) says nothing more than $\{\mathbf{P}^{(1)}\}^{p-m} = \{\mathbf{P}^{(1)}\}^{p-n}\{\mathbf{P}^{(1)}\}^{n-m}$ when the transition probabilities are constant. Its generalization, the **Chapman–Kolmogorov equation**,*,† holds for all Markov processes (discrete and continuous):

$$f_{X(t_2)|X(t_1)}\left(x_2 \mid x_1\right) = \int_{-\infty}^{\infty} f_{X(t_2)|X(t)}\left(x_2 \mid x\right) f_{X(t)|X(t_1)}\left(x \mid x_1\right) dx \quad \left(t_1 < t < t_2\right). \quad (4.71)$$

The Chapman–Kolmogorov equation is established by reasoning as we did for Equation 4.70.

One property of any finite-state transition probability matrix $\mathbf{P}^{(1)}$ is immediate; the sum of the entries in, say, column j is unity because this equals the sum of all the exit probabilities (including stay-at-home) from corner j. (Confirm this for the four-corner transition matrix (4.67).)

An interesting wrinkle for stationary Markov processes is provided by considering the possibility that one of the states is "absorbing," in the sense that all of the exit probabilities from, say, corner j in the four-corner game are zero; one can enter, but never leave, that corner. An absorbing state is evidenced by the appearance of a column in $\mathbf{P}^{(1)}$ consisting of one 1 and the rest 0's. Of course the *a priori* "occupancy probabilities" become fixed after an absorbing state is entered: $\text{Prob}\{X(n) = j\} = 1$, $\text{Prob}\{X(n) = k \neq j\} = 0$.

But even if there are no absorbing states, it may still be possible for a system to settle into an "equilibrium" of sorts where the *a priori* probabilities remain *fixed*, although they are not 0's and 1's. For example, in the four-corner game characterized by Figure 4.5, if one reaches a state where the occupancy probabilities $P(n_0)$ are "just right," the probabilities for subsequent stages are all the same:

$$P\left(n_0 + 1\right) = \mathbf{P}^{(1)}P\left(n_0\right) = \begin{bmatrix} 0.5 & 0.1 & 0 & 0.3 \\ 0.3 & 0.5 & 0.2 & 0 \\ 0 & 0.4 & 0.5 & 0.2 \\ 0.2 & 0 & 0.3 & 0.5 \end{bmatrix}\begin{bmatrix} 0.2037... \\ 0.2407... \\ 0.2963... \\ 0.2593... \end{bmatrix} = \begin{bmatrix} 0.2037... \\ 0.2407... \\ 0.2963... \\ 0.2593... \end{bmatrix} = P\left(n_0\right).$$

And in fact every initial configuration will converge to this equilibrium state because some software experimentation reveals that the powers of the matrix $\mathbf{P}^{(1)}$ converge to a matrix with $P(n_0)$ in each column:

$$\left[\mathbf{P}^{(1)}\right]^n = \begin{bmatrix} 0.5 & 0.3 & 0 & 0.2 \\ 0.2 & 0.5 & 0.3 & 0 \\ 0 & 0.2 & 0.5 & 0.3 \\ 0.3 & 0 & 0.2 & 0.5 \end{bmatrix}^n \rightarrow \begin{bmatrix} 0.2037... & 0.2037... & 0.2037... & 0.2037... \\ 0.2407... & 0.2407... & 0.2407... & 0.2407... \\ 0.2963... & 0.2963... & 0.2963... & 0.2963... \\ 0.2593... & 0.2593... & 0.2593... & 0.2593... \end{bmatrix}.$$

---

* Kolmogoroff, A. 1931. Über die analytischen Methoden in der Wahrscheinlichkeitsrechnung. *Mathematische Annalen* 104: 415–458.
† Chapman, S. 1928. On the Brownian displacements and thermal diffusion of grains suspended in a nonuniform fluid. *Proceedings of the Royal Society of London, Series A* 119: 34–54.

Such a matrix multiplied by any "probability vector" whose components sum to 1 yields $P(n_0)$. (Why?)

The questions naturally arise as to the generality of these observations:

1. Does every finite-state stationary Markov process have equilibrium states?
2. If so, does every initial configuration converge to an equilibrium state?

These are matrix-theory questions because (1) asks if for every possible transition probability matrix—that is, for every square matrix $\mathbf{P}^{(1)}$ with nonnegative entries and all column sums equal to unity—is there a column vector P with nonnegative entries that sum to unity that satisfies

$$\mathbf{P}^{(1)} \bullet P = P. \tag{4.72}$$

and (2) asks if the products

$$\mathbf{P}^{(1)} \mathbf{P}^{(1)} \mathbf{P}^{(1)\cdots} \mathbf{P}^{(1)} P \tag{4.73}$$

converge for all such $\mathbf{P}^{(1)}$ and P.

Equation 4.72 requires that every suitable transition matrix $\mathbf{P}^{(1)}$ has an eigenvector of nonnegative entries with eigenvalue one. Recall that a matrix $\mathbf{A}$ has an eigenvalue λ if and only if $\mathbf{A} - \lambda\mathbf{I}$ is singular—for example, its rows are linearly dependent. So is $\mathbf{P}^{(1)} - (1)\mathbf{I}$ singular?

$$\mathbf{P}^{(1)} - (1)\mathbf{I} = \begin{bmatrix} p_{11} - 1 & p_{12} & p_{13} & p_{14} \\ p_{21} & p_{22} - 1 & p_{23} & p_{24} \\ p_{31} & p_{32} & p_{33} - 1 & p_{34} \\ p_{41} & p_{42} & p_{43} & p_{44} - 1 \end{bmatrix}.$$

The answer is *yes*, because the sum of its rows is $\{0\ 0\ 0\ 0\}$ (recall that each vertical sum in $\mathbf{P}^{(1)}$ is 1). And in fact one can prove that there is an eigenvector for this eigenvalue that is nonnegative. The answer to question (1) is *yes*.

But the answer to question (2) is no. The Markov process described by Figure 4.7 simply cycles the occupancy probabilities for the four corners at each transition; they will not converge unless they *start out* at the equilibrium values $\{1/4, 1/4, 1/4, 1/4\}$.

The transition matrix for this situation is

$$P^{(1)} = \begin{bmatrix} 0 & 0 & 0 & 1 \\ 1 & 0 & 0 & 0 \\ 0 & 1 & 0 & 0 \\ 0 & 0 & 1 & 0 \end{bmatrix}.$$

However, one can prove that if $\mathbf{P}^{(1)}$, or any power of $\mathbf{P}^{(1)}$, contains no zeros, then (2) is true; all states do converge to the equilibrium state.

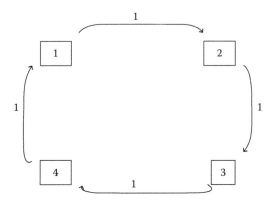

FIGURE 4.7 Cyclic Markov process.

ONLINE SOURCES

Demonstrations of Markov processes:

Boucher, C. Finite-state, discrete-time Markov chains. Wolfram Demonstrations Project. Accessed June 24, 2016. http://demonstrations.wolfram.com/FiniteStateDiscrete TimeMarkovChains/.

Boucher, C. A two-state, discrete-time Markov chain. Wolfram Demonstrations Project. Accessed June 24, 2016. http://demonstrations.wolfram.com/ATwoStateDiscrete TimeMarkovChain.

SUMMARY: COMMON RANDOM PROCESS MODELS

Randomly shifted sine wave: $X(t) = A\cos(\Omega t + \Phi)$, A and Ω nonrandom, $f_\Phi(\varphi) = 1/2\pi$:

$$\mu = 0, \ R(\tau) = A^2\frac{\cos\Omega\tau}{2}, \ S(f) = \frac{\pi A^2}{2}\delta(|2\pi f| - \Omega) = \frac{A^2}{4}\delta\left(|f| - \frac{\Omega}{2\pi}\right)$$

Bernoulli trials: $\text{Prob}\{a\} = p$, $\text{Prob}\{b\} = 1 - p$.

$$\mu_{Bernoulli} \equiv E\{X(n)\} = pa + (1-p)b.$$

$$R_{Bernoulli}(n_1, n_2) \equiv E\{X(n_1)X(n_2)\} = \begin{cases} pa^2 + (1-p)b^2 & \text{if } n_1 = n_2 \\ \mu^2_{Bernoulli} & \text{if } n_1 \neq n_2. \end{cases}$$

$$S_{Bernoulli}(f) = \mu^2_{Bernoulli}\delta(f) + pa^2 + (1-p)b^2 - \mu^2_{Bernoulli}.$$

Binomial process (accumulation of Bernoulli trials):

$$\mu_{binomial}(n) = n\mu_{Bernoulli} = n[pa + (1-p)b].$$

$$R_{binomial}(n_1, n_2) = n_1 n_2 \mu^2_{Bernoulli} + [pa^2 + (1-p)b^2 - \mu^2_{Bernoulli}]\min(n_1, n_2)$$

Poisson process, average λ events per unit time:

$$\mu(t) = \lambda t, \ R(t_1, t_2) = \lambda \min(t_1, t_2) + \lambda^2 t_1 t_2.$$

Random telegraph signal (popcorn noise), average λ events per unit time, levels $\pm a$:

$$\mu = 0, \ R(\tau) = a^2 e^{-2\lambda \tau}, \ S(f) = a^2 \frac{\lambda}{(\pi f)^2 + \lambda^2}.$$

Shot noise (current noise), average λ events per unit time:

$$\mu = \lambda, R_{shot}(\tau) = \lambda \delta(\tau) + \lambda^2, \ S(f) = \lambda^2 \delta(f) + \lambda.$$

Random walk, Prob{step s to the right each time τ} = p:

$$\mu(n) = n(2p-1)s, \ R(n_1, n_2) = s^2 \min(n_1, n_2).$$

Wiener process, $p = 1/2$:

$$\mu = 0, \ R(t_1, t_2) = \alpha \min(t_1, t_2).$$

Brownian motion:

$$\mu = 0, \ R(\tau) = \alpha \delta(\tau), \ S = \alpha.$$

Power Spectral Density Curves (Figure 4.8).

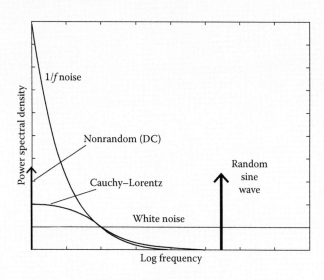

FIGURE 4.8 Power spectral densities.

EXERCISES

SECTION 4.3

1. Show that if every root r of Equation 4.9 satisfies $|r| < 1$ and $X(n)$ is stationary (in the mean), then $E\{X(n)\} = 0$. *Hint*: Derive

$$E\{X(n)\}\left[1-\sum a(k)\right]=0$$

and note that $r = 1$ is not a root.

SECTION 4.4

2. A process $X(n)$ is defined as follows. You start at $X(0) = 0$. Every hour you flip a coin; if it is heads, you move 1 cm to the right; if it is tails, 1 cm to the left. Then you reduce your X-coordinate by 50%. In other words, $X(n) = 0.5\{X(n-1) \pm 1\}$. Set up the Yule–Walker equations that describe this process. What is the autocorrelation $R_X(k)$ for $k = -2, -1, 0, 1, 2$?

3. Show that for the ARMA(1, 0) model

$$R_X(0) = \frac{b(0)^2 \sigma^2}{1-a(1)^2}, \quad R_X(n) = a(1)R_X(n-1) = a(1)^n R_X(0) \ \text{ for } n > 0$$

4. Show that the autocorrelation function for the ARMA (2, 0) process

$$X(n) = a_1 X(n-1) + a_2 X(n-2) + bV(n)$$

is the sum of two decaying exponentials by following these steps;

   a. Show that the general solution to the Yule–Walker equations, for $n \geq 2$, is given by $R_X(n) = c_+ r_+^n + c_- r_-^n$, where the cs are arbitrary constants and $r_\pm = a_1/2 \pm \sqrt{a_1^2 + 4a_2}/2$.

   b. Show that $R_X(0) = b^2 \sigma^2 \dfrac{a_2 - 1}{(a_2 + 1)\left[a_1^2 - (1-a_2)^2\right]}$ and

   $$R_X(1) = b^2 \sigma^2 \frac{-a_1}{(a_2 + 1)\left[a_1^2 - (1-a_2)^2\right]}.$$

   c. Choose the c's so that the formula in (a) matches those in (b). Derive

   $$R_X(n) = \frac{b^2 \sigma^2 \left\{-\left[r_-(a_2 - 1) + a_1\right]r_+^n + \left[r_+(a_2 - 1) + a_1\right]r_-^n\right\}}{(a_2 + 1)\left[a_1^2 - (1-a_2)^2\right]\sqrt{a_1^2 + 4a_2}}.$$

SECTION 4.5

5. Find $a(1)$ and $b(0)$ so that the stationary process satisfying $X(n) = a(1) X(n-1) + b(0) V(n)$, with $V(n)$ a white noise process and $V(n)$ independent of $X(m)$ for $m < n$ and power $\sigma_v^2 = 4$, has an autocorrelation function satisfying $R_X(0) = 4$, $R_X(1) = 3$. For these $a(1)$ and $b(0)$, what is $R_X(2)$?

6. The *estimated* autocorrelations $R_X(k)$ for a random process $X(n)$ for $k = 0,1,2$ are

$$R_X(0) = 2;\ R_X(1) = 1;\ R_X(2) = 1.$$

Estimate the power spectrum of $X(n)$ under each of these assumptions:

   a. $X(n)$ is an AR(2) process.
   b. $X(n)$ is an ARMA(1,1) process.
   c. $X(n)$ is an MA(2) process.

7. What are the coefficients for an ARMA(3,0) process driven by white noise with power 4, whose autocorrelation appears to be $R_X(k) = 6/(k^2 + 1)$? How many of the values of the autocorrelation does your model fit?

SECTION 4.7

8. Show that measurements of $X(t) = A\cos(\Omega t + \varphi)$ at the times $t_0 = -2\tau$, $t_1 = -\tau$, $t_2 = \tau$, $t_3 = 2\tau$ completely determine A, Ω, and φ if $0 < \tau < \pi/\Omega$. (*Hint*: One scheme for extracting the constants is

$$\Omega = \frac{1}{\tau}\arccos\left[\frac{1}{2}\frac{X(t_0) - X(t_3)}{X(t_1) - X(t_2)}\right], A = \sqrt{\left\{\frac{X(t_1) + X(t_2)}{2\cos\Omega\tau}\right\}^2 + \left\{\frac{X(t_1) - X(t_2)}{2\sin\Omega\tau}\right\}^2},$$

$$\varphi = \arcsin\frac{X(t_1) - X(t_2)}{-2A\sin\Omega\tau} = \arccos\frac{X(t_1) + X(t_2)}{-2A\cos\Omega\tau}.)$$

9. Consider a random sine wave $X(t) = \cos(\Omega t + \Theta)$, where Θ has the probability density function $f_\Theta(\theta)$ for $-\pi \le \theta \le \pi$

   a. Show that if $E\{\cos\theta\} = E\{\sin\theta\} = E\{\cos 2\theta\} = E\{\sin 2\theta\} = 0$, then $X(t)$ is wide sense stationary (Equations 3.2 and 3.4).
   b. Give an example of a nonuniform PDF $f_\Theta(\theta)$ satisfying the conditions in part (a).
   c. Assuming (a) is satisfied, determine analytically how the PSD of $X(t)$ depends on $f_\Theta(\theta)$, showing that all the power is concentrated at the angular frequency $(\pm)\Omega$. Give an intuitive explanation for your result.

10. (**Doppler effect**) A random sine wave $\cos(\Omega t + \Theta)$ is transmitted from a fixed antenna and received by a detector moving away from the antenna with velocity v. Θ is a random variable whose pdf satisfies the condition in Problem 9.

   a. Assume that the signal does not attenuate with distance from the antenna (e.g., it is a plane wave). Argue that the received signal has the form $X(t) = \cos[(t - (a + vt)/c) + \Theta]$, where a is the separation at $t = 0$ and c is the speed of wave propagation (also constant).
   b. Use your answers to Problem 9 to describe the PSD of the received signal.

11. Let $p(t)$ be a deterministic periodic waveform with period T. A random process is constructed according to $X(t) = p(t - \Theta)$ where Θ is a random variable uniformly distributed over $[0, T]$
    a. Show that $X(t)$ is wide sense stationary (Equations 3.2 and 3.4).
    b. In the expansion $X(t) = \sum_{n=-\infty}^{\infty} a_n e^{j2\pi nt/T}$ expressing $X(t)$ as a Fourier series, what is the formula for the coefficients $\{a_n\}$ expressed as integrals involving $p(t)$?
    c. Show that the PSD of $X(t)$ is a superposition of delta functions restricted to angular frequencies that are integer multiples of $2\pi/T$.

SECTION 4.9

12. If the time to failure T (years) of an electronic device component follows the exponential distribution $f_T(t) = 2e^{-2t}$, and the cost $C(t)$ of running the device for t years is $75t$, find the PDF of $C(t)$. What is its mean?
13. Suppose that the arrival of messages at a node in a network is a Poisson process with an average arrival rate λ. With probability p, the message contains only data; otherwise, it contains text. What is the pdf of $Y(t)$, the number of data messages arriving by time t?
14. Calculate the mean, autocorrelation, and PSD for a random telegraph signal $X(t)$ (Example 4.1) alternating between the values a and b, instead of $\pm a$.
15. Calculate the mean, autocorrelation, and PSD for a three-level random telegraph signal (Example 4.1) taking, with equal probabilities, the values a, b, and c, instead of $\pm a$.
16. Suppose the random telegraph signal in Example 4.1 arrives at an unreliable receiver that, on a given event, might simply fail to change sign—*staying* at $+a$ or $-a$. The failures occur independently with probability p. Find the mean and autocorrelation of the received signal.
17. The mean and arrival rate of the random telegraph signal described in Problem 14 are to be estimated by sampling the signal $X(t)$ every T seconds. (Of course the values a and b are known.) These samples are averaged to compute an estimate of the mean:

$$\hat{\mu}_X = \frac{1}{n}\sum_{k=1}^{n} X(kT).$$

    a. How you do estimate the arrival rate from this number?
    b. Express the mean squared error (MSE) between the true mean and its estimate $\hat{\mu}_X$.
    c. Compute the MSE for $n = 100$ and $n = 1000$, and the values $\lambda T = 0.5$, 1, and 2. What do you conclude?
18. If $X(t)$ is a Poisson process with arrival rate λ, obviously the "centered" process $Y(t) \equiv X(t) - \lambda t$ has zero mean. What is $R_Y(t_1, t_2)$?

19. How long must one wait to ensure that at least one Poisson event occurs with probability 99%, if the arrival rate is 1 per hour?

20. Tourists arriving at the Fountain of Youth constitute a Poisson process with arrival rate λ. The tourist requires T minutes to bathe in the water; then he or she departs. The Fountain is large enough to accommodate all arrivals; there is no waiting for a place.

    a. What is the probability that there are n tourists at the Fountain?

    b. Now suppose that half the tourists (statistically independently) bathe twice as long. What is the probability that there are 0 tourists at the Fountain?

    c. Under the condition (b), what is the probability that there are n tourists at the Fountain?

21. We are going to test the accuracy of the limiting expression in (4.49). Use software to compare (4.48) with (4.49) for $n = 2$, $t = 20$, $\lambda = 1/3$. Test the accuracy for $m = 10, 100, 1000$, with $T = m/\lambda = 3m$. You will probably need to use Simpson's approximation for the factorial, $p! \approx \sqrt{2\pi p}\left(p/e\right)^{p}$, and to arrange the order of the calculations carefully to avoid overflow on your computer.

22. In Section 4.9, we argued that the waiting time between events for a Poisson process was described by an exponential pdf. Prove the converse: If the probability that the first event after any given time T occurs in the interval $(t, t + dt)$ is $\lambda e^{-\lambda(t-T)}\,dt$ (for $t > T$), then

    a. The probability of 0 events in an interval of length t is $e^{-\lambda t}$

    b. The probability of 1 event in an interval of length t is $\lambda t e^{-\lambda t}$

    c. The probability of 2 events in an interval of length t is $(\lambda t)^2 e^{-\lambda t}/2$

    d. In general, the probability of n events in an interval of length t is given by formula (4.50)

    (*Hint*: Prob{0 events occur in $(0, t)$} = Prob{1st event occurs in $(\tau, \tau + d\tau)$}, summed over $t < \tau < \infty$.

    Prob{1 event occurs in $(0, t)$} = Prob{1st event occurs in $(\tau, \tau + d\tau)$} × Prob{0 events occur in $(\tau + d\tau, t)$}, summed over $0 < \tau < t$.

    Prob{2 events occur in $(0, t)$} = Prob{1st event occurs in $(\tau_1, \tau_1 + d\tau_1)$} × Prob{1st event after that occurs in $(\tau_2, \tau_2 + d\tau_2)$} × Prob{0 events occur in $(\tau_2 + d\tau_2, t)$}, summed over $0 < \tau_1 < t$, $\tau_1 < \tau_2 < t$, and so on.)

SECTION 4.10

23. Define the m-step *incremental* random walk process to be $Y_{RandWalk(m)}(n\tau) = X_{RandWalk}(n\tau) - X_{RandWalk}([n - m]\,\tau)$, where $X_{RandWalk}(n\tau)$ is defined in Section 4.10. Find the mean and autocorrelation of $Y_{RandWalk(m)}(n\tau)$.

24. Generalize the random walk process of Section 4.10 so that the length S of the step taken at the time $n\tau$ is a continuous random variable with pdf $f_S(s)$; the step sizes are independent and identically distributed. Express the mean and autocorrelation of the process in terms of μ_S and σ_S.

Section 4.11

25. If $X(n)$ is the Markov process for the four-corner game described in Section 4.11, discuss its mean and autocorrelation. What are their values after equilibrium has been reached?

26. Consider a two-state Markov chain with a transition probability matrix of the form

$$\mathbf{P}^{(1)} = \begin{bmatrix} p & 1-p \\ 1-p & p \end{bmatrix}.$$

    a. Using a computer, calculate $[\mathbf{P}^{(1)}]^{10}$ for several different values of p (between 0 and 1, of course). What do you see?

    b. Use eigenvector theory to explain what you see. (It is easy to diagonalize $\mathbf{P}^{(1)}$, even for general p.)

27. A random voltage signal is generated as follows. At each switching time, if the voltage level is 0 V, it switches to +1 V with probability p and to −1 V with probability $1 - p$; if the voltage level is *not* zero, it switches back to zero. The starting level is 0 V.

    a. Model this process as a Markov process. Specify the states and the transition probability matrix.

    b. What is the pdf for the voltage at the nth stage?

    c. What are the equilibrium states?

28. A p,n junction initially has s electrons on the n side and s holes on the p side. At each time interval, two random carriers from opposite sides cross the junction. Model this as a Markov process, designating the states by the number of electrons on the p side. Describe the transition probability matrix.

29. Each hour a jar of jelly beans is shaken, and one is selected at random. The selected bean is marked and returned to the jar.

    a. Model this as a Markov process whose states are designated by the number of marked jelly beans in the jar.

    b. If the first time a marked jelly bean is reselected occurs on the nth trial, how would you estimate the number of beans M in the jar? (*Hint:* Choose M to maximize the probability that the first "repeat bean" occurs on the nth trial.)

5 Least Mean-Square Error Predictors

In this and the next chapter we return to the task, introduced in Section 2.3, of prediction for a random process. With the objective of achieving the least mean-square error (LMSE), we survey the techniques for predicting under various circumstances.

The reader should note that we use the same mathematical procedure throughout the two chapters:

- Propose a formula for the predictor.
- Formulate the prediction *error* by subtracting the quantity to be predicted.
- Square and take the expected value to formulate the MSE.
- Minimize the MSE by setting its derivatives, with respect to the parameters, equal to zero.

We emphasize that these predictors entail only second-order statistics—means and covariances. A detailed knowledge of the distributions can enhance the evaluation of their performance but is unnecessary for their construction.

5.1 THE OPTIMAL CONSTANT PREDICTOR

As a starting point, consider the task of choosing a *constant* C to predict the value of a random variable Y. As in Section 2.3, we focus on the error in the prediction, $(Y - C)$, and we choose the value of C that promises the lowest expectation for the squared error—the LMSE.

This value is easy to deduce; we simply express the MSE explicitly and apply the usual second-moment identity $\overline{Z^2} = \overline{Z}^2 + \sigma_Z^2$ to $Z \equiv (Y - C)$ (which has the same standard deviation as Y; see Section 1.8):

$$E\left\{(Y-C)^2\right\} = \left[\text{mean of } (Y-C)\right]^2 + [\text{standard deviation of } (Y-C)]^2 = (\overline{Y} - C)^2 + \sigma_Y^2.$$

Clearly, this is minimized when $C = \overline{Y}$. Thus the LMSE predictor of a random variable Y is (not surprisingly) its mean $\overline{Y}$, and the MSE in this prediction is its variance σ_Y^2. (The *root*-mean-squared [RMS] error is its standard deviation σ_Y.) Also, as we anticipated in Section 2.3, we need to know the mean and standard deviation to implement and assess this predictor.

5.2   THE OPTIMAL CONSTANT-MULTIPLE PREDICTOR

Now, suppose a pair of random variables, X and Y, are related (such as the local temperatures in Boston and New York; recall Section 2.3). Intuitively, this means that if the value that X takes is known, then we have some extra information about the value Y will take. (If X and Y are independent, of course, knowing X tells us nothing about Y, and we revert to the predictor $\bar{Y}$.) How can we exploit the knowledge of X's value to create a better predictor for Y?

First, we propose constructing a simple predictor of the form $\hat{Y} = wX$, a constant multiple of X.

What would be the best such predictor? We employ the LMSE criterion and begin by expressing the MSE:

$$\text{MSE} = E\left\{\left(\hat{Y} - Y\right)^2\right\} = E\left\{(wX - Y)^2\right\} = E\left\{w^2X^2 + Y^2 - 2wXY\right\} = w^2\overline{X^2} + \overline{Y^2} - 2w\overline{XY}.$$

At the MSE's minimum point, its derivative with respect to w is zero:

$$\frac{\partial}{\partial w}E\{(\hat{Y}-Y)^2\} = 2w\overline{X^2} - 2\overline{XY} = 0 \quad \text{or} \quad w = \frac{\overline{XY}}{\overline{X^2}} = \frac{R_{XY}}{\overline{X^2}},$$

and the predictor is

$$\hat{Y} = wX = \frac{R_{XY}}{\overline{X^2}}X. \tag{5.1}$$

Again we need to know the moments to implement this predictor.

5.3   DIGRESSION: ORTHOGONALITY

Recall the properties of the dot product in three-dimensional vector analysis:

$$\vec{v} \cdot \vec{u} = |\vec{v}|\cos\theta\,|\vec{u}|, \quad \vec{v}\cdot\vec{v} = |\vec{v}|^2, \quad \vec{u}\cdot\vec{u} = |\vec{u}|^2, \tag{5.2}$$

where θ is the angle between $\vec{v}$ and $\vec{u}$ and $|\vec{v}|$, $|\vec{u}|$ are the lengths of the vectors.

We can use the dot product to express the orthogonal projection of one vector onto another, as in Figure 5.1.

The length of $\vec{v}_{proj}$ is $|\vec{v}|\cos\theta$; its direction is that of the unit vector $\vec{u}/|\vec{u}|$; thus,

$$\vec{v}_{proj} = |\vec{v}|\cos\theta\frac{\vec{u}}{|\vec{u}|} = \frac{|\vec{v}|\cos\theta|\vec{u}|}{|\vec{u}|}\frac{\vec{u}}{|\vec{u}|} = \frac{\vec{v}\cdot\vec{u}}{\vec{u}\cdot\vec{u}}\vec{u}. \tag{5.3}$$

Now compare the identities (5.2) with the expressions for the second moments for zero-mean random variables (Section 1.10):

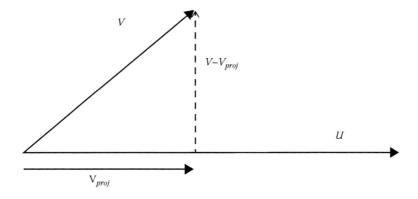

FIGURE 5.1 Orthogonal projection.

$$\vec{v}\cdot\vec{u} = \left|\vec{v}\right|\cos\theta\left|\vec{u}\right|, \quad \vec{v}\cdot\vec{v} = \left|\vec{v}\right|^2, \quad \vec{u}\cdot\vec{u} = \left|\vec{u}\right|^2, \qquad \text{(5.2 repeated)}$$

$$\overline{XY} = \sigma_X\rho\sigma_Y, \qquad \overline{XX} = \sigma_X^2, \qquad \overline{YY} = \sigma_Y^2.$$

The dot products are perfectly analogous to the second moments if we regard σ_X and σ_Y as the "lengths" of X and Y and the correlation coefficient ρ as the cosine of the "angle between X and Y." After all ρ lies between -1 and $+1$, just like the cosine (Section 1.10). In this vein, we say two random variables X and Y are "orthogonal" if $E\{XY\} = 0$ (so the angle is 90°). Note that this nomenclature is only consistent with these analogies when the variables have mean zero. (However, it is often applied in all cases.)

The random-variable analog to the orthogonal projection formula (5.3) is

$$Y_{proj} = "\frac{Y\cdot X}{X\cdot X}X" = \frac{\overline{YX}}{\overline{XX}}X = \frac{R_{XY}}{\overline{X^2}}X,$$

which is identical to the formula in the preceding section for the LMSE predictor for Y of the form wX (Equation 5.1). That is, just as $\vec{v}_{proj}$ is the closest vector to $\vec{v}$ of the form $w\vec{u}$ minimizing the squared distance $\left\|\vec{v} - w\vec{u}\right\|^2$, so $\hat{Y}$ is the "closest" random variable to Y of the form wX minimizing the mean-squared "distance" (i.e., error).

The vector analogy is useful in remembering the LMSE formula. Furthermore, note that $\vec{v} - \vec{v}_{proj}$ is orthogonal to $\vec{v}_{proj}$ in Figure 5.1. By analogy, one would then speculate that the prediction error $Y - \hat{Y}$ is orthogonal (in the statistical sense) to the LMSE predictor $\hat{Y}$:

$$E\left\{\left(Y - \hat{Y}\right)\hat{Y}\right\} = 0. \qquad \text{(5.4)}$$

And in fact we can see this is true by writing it out:

$$E\left\{\left(Y - \hat{Y}\right)\hat{Y}\right\} = E\left\{\left(Y - wX\right)\left(wX\right)\right\} = w\overline{YX} - w^2\overline{X^2} = 0 \quad for \quad w = \frac{\overline{YX}}{\overline{X^2}}.$$

5.4   MULTIVARIATE LMSE PREDICTION: THE NORMAL EQUATIONS

To see how formula for the LMSE predictor $\hat{Y} = wX = \dfrac{\overline{YX}}{\overline{X^2}}X$ generalizes, we now assume there are three related random variables X_1, X_2, and Y, and we wish to construct a predictor for Y that exploits knowledge of the values taken by X_1 and X_2. We propose constructing the linear predictor $\hat{Y}$ in the form

$$\hat{Y} = w_1 X_1 + w_2 X_2, \tag{5.5}$$

with constants w_1 and w_2. What would be the best such predictor?

Again we select the LMSE predictor, the one which minimizes the expected value of the square of the error $w_1 X_1 + w_2 X_2 - Y$. We calculate

$$E\left\{\left(w_1 X_1 + w_2 X_2 - Y\right)^2\right\} = E\left\{w_1^2 X_1^2 + w_2^2 X_2^2 + Y^2 + 2w_1 w_2 X_1 X_2 - 2w_1 X_1 Y - 2w_2 X_2 Y\right\}$$

$$= w_1^2 \overline{X_1^2} + w_2^2 \overline{X_2^2} + \overline{Y^2} + 2w_1 w_2 \overline{X_1 X_2} - 2w_1 \overline{X_1 Y} - 2w_2 \overline{X_2 Y}$$

At the minimum point of the MSE, its partial derivatives with respect to w_1 and w_2 are each zero:

$$\frac{\partial}{\partial w_1} E\left\{\left(w_1 X_1 + w_2 X_2 - Y\right)^2\right\} = 2w_1 \overline{X_1^2} + 2w_2 \overline{X_1 X_2} - 2\overline{X_1 Y} = 0,$$

$$\frac{\partial}{\partial w_2} E\left\{\left(w_1 X_1 + w_2 X_2 - Y\right)^2\right\} = 2w_2 \overline{X_2^2} + 2w_1 \overline{X_1 X_2} - 2\overline{X_2 Y} = 0,$$

or (dropping the 2s)

$$\begin{bmatrix} \overline{X_1^2} & \overline{X_1 X_2} \\ \overline{X_1 X_2} & \overline{X_2^2} \end{bmatrix} \begin{bmatrix} w_1 \\ w_2 \end{bmatrix} = \begin{bmatrix} \overline{X_1 Y} \\ \overline{X_2 Y} \end{bmatrix}, \tag{5.6}$$

whose solution is

$$w_1 = \frac{\overline{X_1 Y}\,\overline{X_2^2} - \overline{X_2 Y}\,\overline{X_1 X_2}}{\overline{X_1^2}\,\overline{X_2^2} - \overline{X_1 X_2}^2}, \quad w_2 = \frac{\overline{X_2 Y}\,\overline{X_1^2} - \overline{X_1 Y}\,\overline{X_1 X_2}}{\overline{X_1^2}\,\overline{X_2^2} - \overline{X_1 X_2}^2}.$$

The generalization is apparent; if X_1, X_2, ..., X_n, and Y are random variables, the coefficients in the LMSE predictor for Y of the form

$$\hat{Y} = w_1 X_1 + w_2 X_2 + \cdots + w_n X_n \tag{5.7}$$

satisfy the system

$$\begin{bmatrix} \overline{X_1^2} & \overline{X_1X_2} & \cdots & \overline{X_1X_n} \\ \overline{X_2X_1} & \overline{X_2^2} & \cdots & \overline{X_2X_n} \\ & & \vdots & \\ \overline{X_nX_1} & \overline{X_nX_2} & \cdots & \overline{X_n^2} \end{bmatrix} \begin{bmatrix} w_1 \\ w_2 \\ \vdots \\ w_n \end{bmatrix} = \begin{bmatrix} \overline{X_1Y} \\ \overline{X_2Y} \\ \vdots \\ \overline{X_nY} \end{bmatrix}. \tag{5.8}$$

And again, we need to know the moments in order to implement this predictor.

Note that the equation for LMSE prediction using a *single* variable X_1 is equivalent to formula (5.1) in Section 5.2. *Note also that we did not make the assumption that the variables were zero-mean, in deriving* (5.8).

Digression: In (deterministic, nonprobabilistic) matrix algebra,* the task of solving a linear system

$$Ax = b \tag{5.9}$$

is straightforward when A is a square matrix with a nonzero determinant. Otherwise (if A is nonsquare or has zero determinant), solutions might not exist, and we have to settle for approximate "least-squares" solutions that minimize the sum-of-squared-differences between the matrix product Ax and the vector b. It is well known that such a least-squares solution x satisfies (exactly) the (square) linear system obtained by multiplying (5.9) on the left by A^T:

$$A^TAx = A^Tb. \tag{5.10}$$

Equations 5.10 are known as the "normal equations"[†] for the problem $Ax \approx b$.

By analogy, then, the equations (5.8) are called the **normal equations** for the coefficients of the LMSE predictor. They can be obtained formally by expressing the prediction "objective" (5.7) as a matrix product

$$\begin{bmatrix} X_1 & X_2 & \cdots & X_n \end{bmatrix} \begin{bmatrix} w_1 \\ w_2 \\ \vdots \\ w_n \end{bmatrix} \approx Y, \tag{5.11}$$

premultiplying by $\begin{bmatrix} X_1 & X_2 & \cdots & X_n \end{bmatrix}^T$, and taking expected values.

---

* Your author's favorite matrix textbook is Saff, E. B. and Snider, A. D. 2015. *Fundamentals of Matrix Analysis*. New York: Wiley.

[†] One would think that the word "normal" must have the longest list of definitions in the dictionary!

5.5   THE BIAS

A predictor $\hat{Y}$ is said to be unbiased if it has the "right" expected value, that is, $E\{\hat{Y}\} = E\{Y\}$. As we have defined it, there is no mechanism in the linear LMSE predictor that enforces this condition. For example, if Y has mean 5 (say), and we try to predict it with *zero*-mean random variables in the form $w_1 X_1 + w_2 X_2$, the bias will be wrong:

$$E\{\hat{Y}\} = w_1 E\{X_1\} + w_2 E\{X_2\} = 0.$$

It would seem (and we will prove) that we will get a better estimator (smaller MSE) if we appropriately adjust the mean of the predictor. And we would expect to be able to accomplish this by including a constant in the predictor, $\hat{Y} = w_1 X_1 + w_2 X_2 + C$, so that C accounts for the bias; that is,

$$C = E\{\hat{Y}\} - w_1 E\{X_1\} - w_2 E\{X_2\}. \tag{5.12}$$

And in fact this is easy to implement using the formalism leading to the normal equations (5.8, Section 5.4). We notationally change our predictor from $\hat{Y} = w_1 X_1 + w_2 X_2$ to $\hat{Y} = w_1 X_1 + w_2 X_2 + w_3 X_3$, where the "random variable" X_3 is the *fixed* number "1."* The normal equations for this bias-corrected LMSE predictor are:

$$\begin{bmatrix} \overline{X_1^2} & \overline{X_1 X_2} & \overline{X_1 X_3} \\ \overline{X_2 X_1} & \overline{X_2^2} & \overline{X_2 X_3} \\ \overline{X_3 X_1} & \overline{X_3 X_2} & \overline{X_3^2} \end{bmatrix} \begin{bmatrix} w_1 \\ w_2 \\ w_3 \end{bmatrix} = \begin{bmatrix} \overline{X_1 Y} \\ \overline{X_2 Y} \\ \overline{X_3 Y} \end{bmatrix},$$

$$\text{or with } X_3 = 1, \quad \begin{bmatrix} \overline{X_1^2} & \overline{X_1 X_2} & \overline{X_1} \\ \overline{X_2 X_1} & \overline{X_2^2} & \overline{X_2} \\ \overline{X_1} & \overline{X_2} & 1 \end{bmatrix} \begin{bmatrix} w_1 \\ w_2 \\ w_3 \end{bmatrix} = \begin{bmatrix} \overline{X_1 Y} \\ \overline{X_2 Y} \\ \overline{Y} \end{bmatrix}.$$

The third equation of this set is $w_1 \overline{X_1} + w_2 \overline{X_2} + w_3 = \overline{Y}$, so comparison with (5.12) reveals that w_3 is the constant (C) that corrects for the bias.[†]

   We also see why correcting for the bias reduces the MSE (or conservatively speaking, never increases it); the *candidates* for the best three-term predictor $\hat{Y} = w_1 X_1 + w_2 X_2 + w_3(1)$ include the two-term predictors $\hat{Y} = w_1 X_1 + w_2 X_2$.

---

* Mean = 1, standard deviation = 0.
[†] Note again that we need to know the first and second moments to implement bias-corrected LMSE predictors.

Should one always use bias-corrected predictors? If we replace the possibly biased predictor $\hat{Y} = w_1X_1 + w_2X_2$ by the unbiased predictor $\hat{Y} = w_1X_1 + w_2X_2 + w_3(1)$, the MSE must be reduced (at least, it will not increase), but this comes at the cost of computing another coefficient for the predictor (with 3, rather than 2, normal equations to be solved). To make a "fair" assessment, we might compare the two-term biased predictor $\hat{Y} = w_1X_1 + w_2X_2$ with a two-term unbiased predictor such as $\hat{Y} = w_1X_1 + w_2(1)$, but it is easy to construct examples that illustrate the superiority both of the former and of the latter:

Example

Let X_1 take the values ±1, each with probability 0.5; and X_2 be ±1 or 0, each with probability 1/3.

If Y just happens to *be* $X_1 + X_2$ (i.e., $w_1 = w_2 = 1$), taking values {−2, −1, 0, 1, or 2}, then of course the predictor $\hat{Y} = X_1 + X_2$ generates NO errors, while any bias-corrected predictor of the form $\hat{Y} = w_1X_1 + w_21$ can take only three values and thus must generate errors.

On the other hand, if Y is the nonrandom *constant* 17, then the bias-corrected predictor $\hat{Y} = 0X_1 + 17$ generates no errors while any predictor of the form $\hat{Y} = w_1X_1 + w_2X_2$ must generate errors since it is not constant.*

As a rule, the extra effort in implementing bias correction by adding a constant to the predictor is not great, so unbiased predictors are generally employed. In many practical applications, all of the random variables $X_1, X_2, \ldots, X_n$, and Y are zero-mean, so bias correction is unnecessary for linear predictors:

$$E\{\hat{Y}\} = w_1 E\{X_1\} + w_2 E\{X_2\} + \cdots w_n E\{X_n\} = 0 = E\{Y\}.$$

ONLINE SOURCES

Demonstrations with biased and unbiased estimators are provided at
Brodie, M. Unbiased and biased estimators. Wolfram Demonstrations Project. Accessed June 24, 2016. http://demonstrations.wolfram.com/UnbiasedAndBiasedEstimators/.

5.6 THE BEST STRAIGHT-LINE PREDICTOR

The LMSE linear predictor of Y in terms of a *single* random variable X that incorporates bias correction,

$$\hat{Y} = wX + C \quad \left(= w_1X_1 + w_2X_2 \text{ with } w_1 = w, X_1 = X, w_2 = C, \text{ and } X_2 \equiv 1\right) \quad (5.13)$$

---

* Actually, $\hat{Y} = 0X_1 + 0X_2$ is constant, but it generates errors.

is called the best *straight-line predictor* because of the familiar form of the equation; *w* equals the slope of the line and *C* is the intercept. These parameters are determined by the normal equations (5.8, Section 5.4):

$$
\begin{bmatrix} \overline{X^2} & \overline{X} \\ \overline{X} & 1 \end{bmatrix} \begin{bmatrix} w_1 \\ C \end{bmatrix} = \begin{bmatrix} \overline{XY} \\ \overline{Y} \end{bmatrix},
$$

whose solution is

$$
w_1 = \frac{\overline{XY} - \overline{X}\,\overline{Y}}{\overline{X^2} - \overline{X}^2} = \frac{\rho \sigma_X \sigma_Y}{\sigma_X^2}, \quad C = \frac{\overline{X^2}\,\overline{Y} - \overline{X}\,\overline{XY}}{\overline{X^2} - \overline{X}^2} = \overline{Y} - w_1 \overline{X}. \tag{5.14}
$$

Digression: A classic (nonrandom) problem in variational calculus is to determine the best straight-line fit $y = mx + b$ to a "scatter diagram" of data points $\{x_1, y_1; x_2, y_2; x_3, y_3; \ldots ; x_k, y_k\}$. Here, "best" means finding slope m and intercept b to minimize the sum of the squares of the differences between y_i and $mx_i + b$. Its solution is well known to be given by

$$
m = \frac{k \sum_{i=1}^{k} x_i y_i - \sum_{i=1}^{k} x_i \sum_{i=1}^{k} y_i}{k \sum_{i=1}^{k} x_i^2 - \left(\sum_{i=1}^{k} x_i \right)^2},
$$

$$
b = \frac{\left(\sum_{i=1}^{k} x_i^2 \right) \left(\sum_{i=1}^{k} y_i \right) - \left(\sum_{i=1}^{k} x_i \right) \left(\sum_{i=1}^{k} x_i y_i \right)}{k \sum_{i=1}^{k} x_i^2 - \left(\sum_{i=1}^{k} x_i \right)^2},
$$

which coincides with the solution (5.14), if we estimate the first and second moments from the data by the usual formulas

$$
\overline{X} \approx \frac{\sum_{i=1}^{k} x_i}{k}, \quad \overline{X^2} \approx \frac{\sum_{i=1}^{k} x_i^2}{k}, \quad \overline{XY} \approx \frac{\sum_{i=1}^{k} x_i y}{k}, \quad \overline{Y} \approx \frac{\sum_{i=1}^{k} y_i}{k}.
$$

Back to the random straight-line predictor. The (least mean-squared) error for the best straight-line predictor, itself, is given by

$$
E\left\{ (w_1 X + C - Y)^2 \right\} = E\left\{ w_1^2 X^2 + C^2 + Y^2 + 2w_1 CX - 2w_1 XY - 2CY \right\}
$$

$$
= w_1^2 \overline{X^2} + C^2 + \overline{Y^2} + 2w_1 C \overline{X} - 2w_1 \overline{XY} - 2C\overline{Y} = \left(1 - \rho^2 \right) \sigma_Y^2 \tag{5.15}
$$

(after insertion of (5.14)).

Note that if the correlation coefficient ρ is one, the expected squared error is zero; the random variable Y is, in fact, exactly $\frac{\rho\sigma_Y}{\sigma_X} X$.* Otherwise, the knowledge of the value of the variable X reduces the uncertainty in Y (i.e., σ_Y) by the factor $\sqrt{1-\rho^2}$.

5.7   PREDICTION FOR A RANDOM PROCESS

What do the normal equations (5.8, Section 5.4) tell us when we apply them to predict the outcomes of a random process? A typical problem is as follows: given measurements of the values of $X(1)$, $X(2)$, ..., $X(n)$, construct the best (LMSE) predictor $\hat{Y}$ of $X(n+1)$ of the form

$$\hat{Y} = \begin{bmatrix} X(n) & X(n-1) & \cdots & X(1) \end{bmatrix} \begin{bmatrix} w_1 \\ w_2 \\ \vdots \\ w_n \end{bmatrix}. \tag{5.16}$$

(It is traditional in this context to write the predictor in "filter" or "autoregressive" form, with the summation index ascending for the coefficients w_i and descending for the data $X(i)$.) We form the normal equations by multiplying (5.16) by $\begin{bmatrix} X(n) & X(n-1) & \cdots & X(1) \end{bmatrix}^T$ and taking expected values. In the normal equations, the second moments now have the interpretation as autocorrelations: $X(i)X(j) \equiv R_X(i, j)$. Therefore the equations take the form

$$\begin{bmatrix} R_X(n,n) & R_X(n,n-1) & \cdots & R_X(n,1) \\ R_X(n-1,n) & R_X(n-1,n-1) & \cdots & R_X(n-1,1) \\ & \vdots & & \\ R_X(1,n) & R_X(1,n-1) & \cdots & R_X(1,1) \end{bmatrix} \begin{bmatrix} w_1 \\ w_2 \\ \vdots \\ w_n \end{bmatrix} = \begin{bmatrix} R_X(n,n+1) \\ R_X(n-1,n+1) \\ \vdots \\ R_X(1,n+1) \end{bmatrix}. \tag{5.17}$$

If the process is stationary, this simplifies to

$$\begin{bmatrix} R_X(0) & R_X(1) & R_X(2) & \cdots & R_X(n-2) & R_X(n-1) \\ R_X(1) & R_X(0) & R_X(1) & \cdots & R_X(n-3) & R_X(n-2) \\ R_X(2) & R_X(1) & R_X(0) & \cdots & R_X(n-4) & R_X(n-3) \\ & & \vdots & & & \\ R_X(n-1) & R_X(n-2) & R_X(n-3) & \cdots & R_X(1) & R_X(0) \end{bmatrix} \begin{bmatrix} w_1 \\ w_2 \\ w_3 \\ \vdots \\ w_{n-1} \\ w_n \end{bmatrix} = \begin{bmatrix} R_X(1) \\ R_X(2) \\ R_X(3) \\ \vdots \\ R_X(n-1) \\ R_X(n) \end{bmatrix}. \tag{5.18}$$

---

* With probability one.

Of course, if the process is stationary, the same coefficients $\{w_1, w_2, \ldots, w_n\}$ are used to predict $X(m + n)$ from $X(m)$, $X(m + 1)$, $X(m + 2)$, $\ldots$, $X(m + n - 1)$, for every m. To predict any value of X from the five preceding values, for example, calculate the coefficients by solving

$$\begin{bmatrix} R_X(0) & R_X(1) & R_X(2) & R_X(3) & R_X(4) \\ R_X(1) & R_X(0) & R_X(1) & R_X(2) & R_X(3) \\ R_X(2) & R_X(1) & R_X(0) & R_X(1) & R_X(2) \\ R_X(3) & R_X(2) & R_X(1) & R_X(0) & R_X(1) \\ R_X(4) & R_X(3) & R_X(2) & R_X(1) & R_X(0) \end{bmatrix} \begin{bmatrix} w_1 \\ w_2 \\ w_3 \\ w_4 \\ w_5 \end{bmatrix} = \begin{bmatrix} R_X(1) \\ R_X(2) \\ R_X(3) \\ R_X(4) \\ R_X(5) \end{bmatrix}. \qquad (5.19)$$

Note the structure of the coefficient matrix; it is constant along diagonals. Such a matrix is called "Toeplitz," and Norman Levinson* developed a fast algorithm for solving these systems.

5.8 INTERPOLATION, SMOOTHING, EXTRAPOLATION, AND BACK-PREDICTION

Because prediction is the most common task confronting a statistician, we have expressed our discussion of LMSE considerations in this context. "Prediction" is really too specific a term for what we have accomplished, however. The variable to be "predicted" could be an *intermediate* value in a random process; perhaps the value was omitted from a measurement. So we might want to construct the LMSE **interpolator** $\hat{Y}$ for $X(n)$ from preceding *and subsequent* values of X. The normal equations (5.8, Section 5.4) for the LMSE interpolator for $X(n)$ of the form $\hat{Y} = w_1 X(n-1) + w_2 X(n+1)$ would read as follows:

$$\begin{bmatrix} R_X(n-1, n-1) & R_X(n-1, n+1) \\ R_X(n+1, n-1) & R_X(n+1, n+1) \end{bmatrix} \begin{bmatrix} w_1 \\ w_2 \end{bmatrix} = \begin{bmatrix} R_X(n-1, n) \\ R_X(n+1, n) \end{bmatrix}$$

or, for a stationary process,

$$\begin{bmatrix} R_X(0) & R_X(2) \\ R_X(2) & R_X(0) \end{bmatrix} \begin{bmatrix} w_1 \\ w_2 \end{bmatrix} = \begin{bmatrix} R_X(1) \\ R_X(1) \end{bmatrix}.$$

We might even want to *improve a measured value* for $X(n)$, using preceding and subsequent values *and X(n) itself*: $\hat{Y} = w_1 X(n-1) + w_2 X(n) + w_3 X(n+1)$. The normal equations for such a **smoothing** operation would look like

---

* Levinson, N. 1947. The Wiener RMS error criterion in filter design and prediction. *J. Math. Phys.* 25:261–278.

$$\begin{bmatrix} R_X(n-1,n-1) & R_X(n-1,n) & R_X(n-1,n+1) \\ R_X(n,n-1) & R_X(n,n) & R_X(n,n+1) \\ R_X(n+1,n-1) & R_X(n+1,n) & R_X(n+1,n+1) \end{bmatrix} \begin{bmatrix} w_1 \\ w_2 \\ w_3 \end{bmatrix} = \begin{bmatrix} R_X(n-1,n) \\ R_X(n,n) \\ R_X(n+1,n) \end{bmatrix}.$$

Similarly, to **extrapolate** a remote future value $X(n+m)$ as a linear combination $w_1 X(n) + w_2 X(n-1) + w_3 X(n-2)$ for a stationary process, use coefficients satisfying

$$\begin{bmatrix} R_X(0) & R_X(1) & R_X(2) \\ R_X(1) & R_X(0) & R_X(1) \\ R_X(2) & R_X(1) & R_X(0) \end{bmatrix} \begin{bmatrix} w_1 \\ w_2 \\ w_3 \end{bmatrix} = \begin{bmatrix} R_X(m) \\ R_X(m+1) \\ R_X(m+2) \end{bmatrix};$$

and to **back-predict** $X(n)$ with $w_1 X(n+1) + w_2 X(n+2)$ use coefficients satisfying

$$\begin{bmatrix} R_X(0) & R_X(1) \\ R_X(1) & R_X(0) \end{bmatrix} \begin{bmatrix} w_1 \\ w_2 \end{bmatrix} = \begin{bmatrix} R_X(1) \\ R_X(2) \end{bmatrix}$$

(Compare with Equations 5.19; for a stationary process back-prediction is the same as forward-prediction).

5.9   THE WIENER FILTER

Now let us consider a generalization of our problem due to Norbert Wiener.* Again consider three random variables, X_1, X_2, and Y, each assumed zero-mean with known autocorrelations and cross-correlations. But now, instead of having exact values for X_1 and X_2, we have "noisy" measurements Z_1 and Z_2; that is,

$$Z_1 = X_1 + e_1, \quad Z_2 = X_2 + e_2 \tag{5.20}$$

where the "measurement errors" e_1 and e_2 are zero-mean random variables uncorrelated with each other and with X_1 and X_2. We seek the best predictor for Y in terms of the corrupted data Z_1 and Z_2:

$$\hat{Y} = w_1 Z_1 + w_2 Z_2 \tag{5.21}$$

---

* Wiener, N. (1949). *Extrapolation, Interpolation, and Smoothing of Stationary Time Series.* New York: Wiley. The predictor is sometimes called the "Wiener–Hopf filter" due to Wiener's collaboration with Eberhard Hopf. Their most famous joint work was Wiener, N. and Hopf, E. 1931. Ueber eine Klasse singulärer Integralgleichungen. *Sitzungber. Akad. Wiss. Berlin*: 696–706. It gave rise to the Wiener–Hopf technique for integral equations, which is used in electromagnetics.

The coefficients must, as always, satisfy the normal equations (5.8, Section 5.4), with Z_1, Z_2 instead of X_1, X_2 :

$$\begin{bmatrix} \overline{Z_1^2} & \overline{Z_1 Z_2} \\ \overline{Z_1 Z_2} & \overline{Z_2^2} \end{bmatrix} \begin{bmatrix} w_1 \\ w_2 \end{bmatrix} = \begin{bmatrix} \overline{Z_1 Y} \\ \overline{Z_2 Y} \end{bmatrix}. \tag{5.22}$$

The correlations in Z_1, Z_2 can easily be reduced to the correlations in X_1, X_2 by using (5.20) and the assumption that the errors are zero-mean and uncorrelated:

$$\overline{Z_1^2} = E\left\{ (X_1 + e_1)^2 \right\} = \overline{X_1^2 + 2e_1 X_1 + e_1^2} = \overline{X_1^2} + 2\overline{e_1 X_1} + \overline{e_1^2} = \sigma_{X_1}^2 + 0 + \sigma_{e_1}^2,$$

$$\overline{Z_2^2} = \sigma_{X_2}^2 + \sigma_{e_2}^2,$$

$$\overline{Z_1 Z_2} = E\left\{ (X_1 + e_1)(X_2 + e_2) \right\} = \overline{X_1 X_2 + e_1 X_2 + e_2 X_1 + e_1 e_2} = \overline{X_1 X_2} + 0 + 0 + 0,$$

$$\overline{Z_1 Y} = E\left\{ (X_1 + e_1)(Y) \right\} = \overline{X_1 Y + e_1 Y} = \overline{X_1 Y} + 0, \text{ and similarly } \overline{Z_2 Y} = \overline{X_2 Y}. \tag{5.23}$$

Thus the normal equations for this predictor are

$$\begin{bmatrix} \sigma_{X_1}^2 + \sigma_{e_1}^2 & \overline{X_1 X_2} \\ \overline{X_1 X_2} & \sigma_{X_2}^2 + \sigma_{e_2}^2 \end{bmatrix} \begin{bmatrix} w_1 \\ w_2 \end{bmatrix} = \begin{bmatrix} \overline{X_1 Y} \\ \overline{X_2 Y} \end{bmatrix} \tag{5.24}$$

with the obvious generalization

$$\left(\begin{bmatrix} \sigma_{X_1}^2 & \overline{X_1 X_2} & \cdots & \overline{X_1 X_p} \\ \overline{X_2 X_1} & \sigma_{X_2}^2 & \cdots & \overline{X_2 X_p} \\ & & \vdots & \\ \overline{X_p X_1} & \overline{X_p X_2} & \cdots & \sigma_{X_p}^2 \end{bmatrix} + \begin{bmatrix} \sigma_{e_1}^2 & 0 & \cdots & 0 \\ 0 & \sigma_{e_2}^2 & \cdots & 0 \\ & & \vdots & \\ 0 & 0 & \cdots & \sigma_{e_p}^2 \end{bmatrix} \right) \begin{bmatrix} w_1 \\ w_2 \\ \vdots \\ w_p \end{bmatrix} = \begin{bmatrix} \overline{X_1 Y} \\ \overline{X_2 Y} \\ \vdots \\ \overline{X_p Y} \end{bmatrix}. \tag{5.25}$$

When X_1, X_2, ..., X_p and Y are consecutive members of a zero-mean stationary random process $X(n)$, corrupted by uncorrelated measurement noise to produce $Z(n) = X(n) + e(n)$, and the measurement errors are identically distributed, the predictor of $Y \equiv X(n+1)$ is called a **Wiener filter**, and it is written in the "filter" form as

$$\hat{Y} = w_1 Z(n) + w_2 Z(n-1) + \cdots + w_p Z(n-p+1). \tag{5.26}$$

Equations 5.25 then become

$$
\left(
\begin{bmatrix}
R_X(0) & R_X(1) & R_X(2) & R_X(3) & \cdots & R_X(p-1) \\
R_X(1) & R_X(0) & R_X(1) & R_X(2) & \cdots & R_X(p-2) \\
& & & \vdots & & \\
R_X(p-1) & R_X(p-2) & R_X(p-3) & R_X(p-4) & \cdots & R_X(0)
\end{bmatrix}
+ \sigma_e^2
\begin{bmatrix}
1 & 0 & \cdots & 0 \\
0 & 1 & \cdots & 0 \\
& & \vdots & \\
0 & 0 & \cdots & 1
\end{bmatrix}
\right)
$$

$$
\times
\begin{bmatrix}
w_1 \\
w_2 \\
\vdots \\
w_p
\end{bmatrix}
=
\begin{bmatrix}
R_X(1) \\
R_X(2) \\
\vdots \\
R_X(p)
\end{bmatrix}. \tag{5.27}
$$

To implement the Wiener filter, we need to know, yet again, the moments for the process X and the noise e.

It is instructive to toy with the notion of an infinite-length Wiener interpolator. The objective would be to find the optimal estimator for, say, $X(0)$ in the form $Y = \sum_{j=-\infty}^{\infty} w(j)Z(-j)$. The optimal estimator for $X(k)$ would then be $\sum_{j=-\infty}^{\infty} w(j)Z(k-j)$, by stationarity. Writing it out in matrix form, we have the estimation problem

$$
\begin{bmatrix} \cdots & Z(2) & Z(1) & Z(0) & Z(-1) & Z(-2) & \cdots \end{bmatrix}
\begin{bmatrix}
\vdots \\
w(-2) \\
w(-1) \\
w(0) \\
w(1) \\
w(2) \\
\vdots
\end{bmatrix}
\approx X(0). \tag{5.28}
$$

This has the format of Equation 5.11 in Section 5.3, so we follow the prescription there for obtaining the normal equations—multiplying (5.28) by $\begin{bmatrix} \cdots & Z(2) & Z(1) & Z(0) & Z(-1) & Z(-2) & \cdots \end{bmatrix}^T$ and taking expected values:

$$\begin{bmatrix} \ddots & \vdots & \vdots & \vdots & \vdots & \vdots & \cdot^{\cdot^{\cdot}} \\ \cdots & \overline{Z(2)^2} & \overline{Z(2)Z(1)} & \overline{Z(2)Z(0)} & \overline{Z(2)Z(-1)} & \overline{Z(2)Z(-2)} & \cdots \\ \cdots & \overline{Z(1)Z(2)} & \overline{Z(1)^2} & \overline{Z(1)Z(0)} & \overline{Z(1)Z(-1)} & \overline{Z(1)Z(-2)} & \cdots \\ \cdots & \overline{Z(0)Z(2)} & \overline{Z(0)Z(1)} & \overline{Z(0)^2} & \overline{Z(0)Z(-1)} & \overline{Z(0)Z(-2)} & \cdots \\ \cdots & \overline{Z(-1)Z(2)} & \overline{Z(-1)Z(1)} & \overline{Z(-1)Z(0)} & \overline{Z(-1)^2} & \overline{Z(-1)Z(-2)} & \cdots \\ \cdots & \overline{Z(-2)Z(2)} & \overline{Z(-2)Z(1)} & \overline{Z(-2)Z(0)} & \overline{Z(-2)Z(-1)} & \overline{Z(-2)^2} & \cdots \\ \cdot^{\cdot^{\cdot}} & \vdots & \vdots & \vdots & \vdots & \vdots & \ddots \end{bmatrix}$$

$$\times \begin{bmatrix} \vdots \\ w(-2) \\ w(-1) \\ w(0) \\ w(1) \\ w(2) \\ \vdots \end{bmatrix} \approx \begin{bmatrix} \vdots \\ \overline{Z(2)X(0)} \\ \overline{Z(1)X(0)} \\ \overline{Z(0)X(0)} \\ \overline{Z(-1)X(0)} \\ \overline{Z(-2)X(0)} \\ \vdots \end{bmatrix}. \tag{5.29}$$

As with equations (5.23),

$$E\{Z(i)Z(j)\} = E\{[X(i)+e(i)][X(j)+e(j)]\}$$
$$= E\{X(i)X(j)\} + E\{e(i)e(j)\} + E\{X(i)e(j)\} + E\{e(i)X(j)\}.$$

We assume the noise is zero-mean and uncorrelated with the process, so the last two terms are zero. The first two are autocorrelations, $R_X(i-j)$ and $R_e(i-j)$; for the sake of generality, we make no assumptions about the noise autocorrelation R_e. Also,

$$E\{Z(i)X(0)\} = E\{X(i)X(0)\} + E\{e(i)X(0)\} = R_X(i)+(0).$$

Observing that autocorrelations are even functions, we can express the normal equations in a suggestive manner:

$$\begin{bmatrix} \ddots & \vdots & \vdots & \vdots & \vdots & \vdots & \vdots & \iddots \\ \cdots & R_X(0)+R_e(0) & R_X(-1)+R_e(-1) & R_X(-2)+R_e(-2) & R_X(-3)+R_e(-3) & R_X(-4)+R_e(-4) & \cdots \\ \cdots & R_X(1)+R_e(1) & R_X(0)+R_e(0) & R_X(-1)+R_e(-1) & R_X(-2)+R_e(-2) & R_X(-3)+R_e(-3) & \cdots \\ \cdots & R_X(2)+R_e(2) & R_X(1)+R_e(1) & R_X(0)+R_e(0) & R_X(-1)+R_e(-1) & R_X(-2)+R_e(-2) & \cdots \\ \cdots & R_X(3)+R_e(3) & R_X(2)+R_e(2) & R_X(1)+R_e(1) & R_X(0)+R_e(0) & R_X(-1)+R_e(-1) & \cdots \\ \cdots & R_X(4)+R_e(4) & R_X(3)+R_e(3) & R_X(2)+R_e(2) & R_X(1)+R_e(1) & R_X(0)+R_e(0) & \cdots \\ \iddots & \vdots & \vdots & \vdots & \vdots & \vdots & \ddots \end{bmatrix}$$

$$\times \begin{bmatrix} \vdots \\ w(-2) \\ w(-1) \\ w(0) \\ w(1) \\ w(2) \\ \vdots \end{bmatrix} = \begin{bmatrix} \vdots \\ R_X(-2) \\ R_X(-1) \\ R_X(0) \\ R_X(1) \\ R_X(2) \\ \vdots \end{bmatrix}. \tag{5.30}$$

Now, it is clear that, row by row, (5.29) expresses the convolution relation:

$$R_X(n) = \sum_{p=-\infty}^{\infty} \left[R_X(n-p) + R_e(n-p) \right] w(p).$$

Therefore, we apply the Fourier convolution theorem (3.9, 3.10) of Section 3.3 and identify the resulting transforms as power spectral densities (Wiener–Khintchine theory, Section 3.5 Equation 3.23):

$$\left[S_X(f) + S_e(f) \right] W(f) = S_X(f), \text{ or } W(f) = \frac{S_X(f)}{S_X(f) + S_e(f)}, \tag{5.31}$$

where $W(f)$ is the Fourier transform of the Wiener coefficients $\{w_i\}$. *The optimal infinite-length linear interpolator can be found by taking the inverse Fourier transform of the ratio of the signal PSD to the sum of the signal and noise PSDs.*

The so-called **Wiener deconvolution**, implemented through Fourier transforms as indicated, has been very successful in such applications as clarifying audio recordings and photographs (using a two-dimensional version); see the sources in the following. Note that for white noise $R_e(k) = \sigma_e^2 \delta_{k0}$, $S_e(f) = \sigma_e^2$, and the normal equations 5.30 resemble 5.27.

ONLINE SOURCES

Many examples and simulations of the Wiener filter are given in

Monique P. F. FIR filtering results review & practical applications. Naval Postgraduate School. Accessed June 24, 2016. http://faculty.nps.edu/fargues/teaching/ec4440/EC4440-00-DL.pdf.

Code and case studies for Wiener deconvolution can be seen at

Noise removal. MathWorks. Accessed June 24, 2016. http://www.mathworks.com/help/images/noise-removal.html.

Deblurring images using a Wiener filter. MathWorks. Accessed June 24, 2016. http://www.mathworks.com/help/images/examples/deblurring-images-using-a-wiener-filter.html.

Scalart, P. Wiener filter for noise reduction and speech enhancement. MathWorks. Accessed June 24, 2016. http://www.mathworks.com/matlabcentral/fileexchange/24462-wiener-filter-for-noise-reduction-and-speech-enhancement/content/WienerNoiseReduction.m.

Malysa, G. Wiener filter demonstration. YouTubeMathWorks. Accessed June 24, 2016. https://www.youtube.com/watch?v=U71siegOD48 [plays a noisy, then filtered, speech by Albert Einstein.]

Bovik, A. and Rajashekar, U. The (1D) digital signal processing gallery. SIVA—DSP. Accessed June 24, 2016. http://live.ece.utexas.edu/class/siva/siva_dsp/siva_dsp.htm [has downloadable audio files and filter codes.]

EXERCISES

SECTION 5.2

1. Refer to the process in Problem 2 of the Exercises for Chapter 4. If one wishes to predict your position $X(n + 1)$ by using a multiple of $X(n)$, what multiple should one use to have the estimator with the LMSE? If $X(5) = 4$, what is the predicted value of $X(6)$?

SECTION 5.4

2. A zero-mean stationary discrete process $X(n)$ has an autocorrelation function given by the following table:

m	0	1	2	3	4
$E\{X(n)X(m+n)\}$	4	3	2	1	0

Find the coefficients in a linear filter that predicts $X(n)$ from the previous three values.

SECTION 5.8

3. You have 1024 samples of an ergodic (stationary) zero-mean random process $X(n)$. By using partitioning and Fourier Transforming, you deduce that the autocorrelation function for X is $1/2^k$, and you wish to predict the value of $X(n + 2)$ as a combination of $X(n)$, $X(n - 1)$, and $X(n - 2)$. The predictor is to

be written as $w(0)X(n)+w(1)\cdot X(n-1)+w(2)X(n-2)$. Find the values of $w(0)$, $w(1)$, $w(2)$ minimizing the mean squared error of the predictions.

4. Show that the optimal linear interpolator for the random telegraph signal (Section 4.9) using two future samples and two past samples depends only on the immediate neighboring samples.

SECTION 5.9

5. A zero-mean stationary discrete random process $X(n)$ has an autocorrelation function $R_X(k) = 10e^{-k}$. A noisy measurement of each $X(n)$ is made, $Z(n) = X(n) + V(n)$, where V is zero-mean white noise, independent of X, with power 0.5. You need to estimate $X(100)$, but the only measurement data available to you are $Z(98)$ and $Z(101)$. Construct the LMSE estimator of $X(100)$ as a linear combination of the subsequent measurement $Z(101)$ and the second day's previous measurement $Z(98)$. If $Z(98) = 10$ and $Z(101) = 11$, what is the estimate for $X(100)$?

6. $X(n)$ is an ARMA(1,1) random process: $X(n) = a(1)X(n-1) + V(n) + b(1)V(n-1)$, where $V(n)$ is white noise with mean 0 and variance σ_V^2.

   a. Design a Wiener filter $\hat{X}(n) = w(1)X(n-1) + w(2)X(n-2)$ that minimizes the mean-square error in the prediction of $X(n)$, and find the minimum MSE.

   b. Consider a predictor of the form $\hat{X}(n) = w(0) + w(1)X(n-1) + w(2)X(n-2)$. Find the values for $w(0)$, $w(1)$, and $w(2)$ that minimize the MSE.

   c. Compare the MSEs of these two predictors.

6 The Kalman Filter

The main innovation of the Wiener filter described in Section 5.9 was that it constructed an optimal estimator of a random variable, not from *values* of related random variables, but from *noisy measurements* of those variables. It demonstrated how to take advantage of knowledge provided by measurements, even when the measurements are not completely reliable. In 1960, Rudolph Kalman* published an even more robust filter, again based on least mean squared error (LMSE) considerations, incorporating noisy system dynamics into the formalism. (So the system is decidedly nonstationary.)

Because the Kalman predictor/filter has proven to be such a valuable tool in engineering, we shall proceed leisurely with the exposition. We first introduce the filter in a very simple context—estimating a constant from two noisy measurements. We then expand the algebraic complexity in stages by generalizing to systems with transitions, noisy dynamical systems, and finally the full matrix formulation that Kalman devised for analyzing the trajectories of moving objects in space.

6.1 THE BASIC KALMAN FILTER

Suppose we have two imperfect measurements, X_{old} and X_{new}, of a number X_{true}. (Two different jewelers weigh a gold ring, say.) We wish to construct the "best" estimate X_{Kalman} that can be formed as a linear combination of X_{old} and X_{new}. We will use the traditional notation in this chapter:

$$X_{Kalman} = KX_{new} + LX_{old}. \tag{6.1}$$

The Kalman formalism assumes that X_{old} and X_{new} are unbiased estimates of the true values (remember that E denotes expected value),

$$E\{X_{old}\} = X_{true}, \qquad E\{X_{new}\} = X_{true},$$

and that the errors e_{old} and e_{new} in these measurements

$$X_{old} - X_{true} = e_{old}, \qquad X_{new} - X_{true} = e_{new}$$

are independent

$$E\{e_{old}\,e_{new}\} = E\{e_{old}\}E\{e_{new}\} = 0 \cdot 0 = 0.$$

---

* Kalman, R.E. 1960. A new approach to linear filtering and prediction problems. *Journal of Basic Engineering* 82: 35–45.

(The errors are zero-mean since the estimates X_{old} and X_{new} of X_{true} are unbiased.) However, one estimate may be more reliable than the other, so σ_{old} and σ_{new} could be different.

We stipulate that the *Kalman estimator* (6.1) should be unbiased and have LMSE. With this formulation "unbiased" implies $K + L = 1$ since

$$X_{true} = E\{X_{Kalman}\} = E\{K X_{new} + L X_{old}\}$$
$$= KE\{X_{new}\} + LE\{X_{old}\} = KX_{true} + LX_{true}.$$

As a result, we can write the Kalman estimator in terms of only one coefficient:

$$X_{Kalman} = KX_{new} + (1 - K) X_{old} \tag{6.2}$$

or, as it is traditionally written

$$X_{Kalman} = X_{old} + K\{X_{new} - X_{old}\}. \tag{6.3}$$

The latter form suggests the interpretation that X_{Kalman} is an update of the estimate X_{old}, taking into account the "improvement" $\{X_{new} - X_{old}\}$ afforded by the new measurement, weighted by the *Kalman gain K*.

The "LSME" dictates, as always, that the mean of the squared error in X_{Kalman}, $E\{[X_{Kalman} - X_{true}]^2\}$, should be minimized as a function of K. Since the formulation (6.1) and (6.2) is quite different from the normal-equations approach (the variables are not zero-mean), we shall work through a new derivation. (But the basic LMSE logic is maintained.) To facilitate this, we express the Kalman error in terms of the estimator errors e_{old} and e_{new} by referencing every estimator in the equation to its true value. This is accomplished by subtracting the mathematical identity

$$X_{true} = KX_{true} + (1 - K) X_{true} \tag{6.4}$$

from (6.2):

$$X_{Kalman} - X_{true} = K\left[X_{new} - X_{true} \right] + (1 - K)\left[X_{old} - X_{true} \right] = Ke_{new} + (1 - K)e_{old}. \tag{6.5}$$

The mean squared error in the Kalman estimator is then

$$E\left\{ \left[X_{Kalman} - X_{true} \right]^2 \right\} = K\left\{ \left[Ke_{new} + (1 - K)e_{old} \right]^2 \right\}$$
$$= K^2 E\{e_{new}^2\} + 2K(1 - K)E\{e_{new}e_{old}\} + (1 - K)^2 E\{e_{old}^2\}$$
$$= K^2 \sigma_{new}^2 + (0) + (1 - K)^2 \sigma_{old}^2. \tag{6.6}$$

The value of K that minimizes the mean squared error is found by setting the derivative equal to zero.

$$\frac{d}{dK} E\left\{\left[X_{Kalman} - X_{true}\right]^2\right\} = 2K\sigma_{new}^2 - 2(1-K)\sigma_{old}^2 = 0, \quad \text{or} \quad K_{Kalman} = \frac{\sigma_{old}^2}{\sigma_{new}^2 + \sigma_{old}^2}.$$

$$(6.7)$$

This is a sensible result. If the error in the old estimate is smaller, K_{Kalman} is less than 1/2, and in $\{K_{Kalman}X_{new} + (1 - K_{Kalman})X_{old}\}$ (6.2) we weigh the old estimate more. And conversely.

Inserting (6.7) into (6.6), we find that the LMSE itself equals the Kalman gain times the mean squared error in the new estimate:

$$\sigma_{Kalman}^2 = E\left\{\left[X_{Kalman} - X_{true}\right]^2\right\} = \left\{\frac{\sigma_{old}^2}{\sigma_{new}^2 + \sigma_{old}^2}\right\}^2 \sigma_{new}^2 + \left\{1 - \frac{\sigma_{old}^2}{\sigma_{new}^2 + \sigma_{old}^2}\right\}^2 \sigma_{old}^2$$

$$= \frac{\sigma_{old}^4}{\left(\sigma_{new}^2 + \sigma_{old}^2\right)^2} \sigma_{new}^2 + \frac{\sigma_{new}^4}{\left(\sigma_{new}^2 + \sigma_{old}^2\right)^2} \sigma_{old}^2 = \frac{\sigma_{old}^2}{\sigma_{new}^2 + \sigma_{old}^2} \sigma_{new}^2 = K_{Kalman}\sigma_{new}^2.$$

$$(6.8)$$

6.2 KALMAN FILTER WITH TRANSITION: MODEL AND EXAMPLES

The Kalman predictor has found wide applicability in dynamic systems, where X undergoes transitions between the measurements of X_{old} and X_{new}. The basic Kalman model for the transition is simple, namely, the new value is simply a multiple of the old value plus a shift

$$X_{new}^{true} = AX_{old}^{true} + B.$$

$$(6.9)$$

Actually, we shall see shortly that to accommodate a larger class of applications, it is advantageous to regard X as a multiple-component *vector*, so that the transition is effected by multiplication by a *matrix A*. The shift B then is also a vector.

Moreover, in many cases the transition is corrupted by noise, leading to the further generalization

$$X_{new}^{true} = AX_{old}^{true} + B + e_{trans}.$$

$$(6.10)$$

Finally, we have to allow the possibility that not all of the components of X_{new}^{true} are measured; or that some are measured more than once, by independent agents; or indeed that the components are not measured directly, but rather in certain combinations. And, of course, the measurements contain noise also. Thus, we propose that the measurement result(s) S are also described by a matrix product

$$S = DX_{new}^{true} + e_{meas} = \begin{bmatrix} d_{11} & d_{12} & \cdots & d_{1p} \\ d_{21} & d_{22} & \cdots & d_{2p} \\ & & \ddots & \\ d_{r1} & d_{r2} & \cdots & d_{rp} \end{bmatrix} \begin{bmatrix} X_1 \\ X_2 \\ \vdots \\ X_p \end{bmatrix}_{new,true} + \begin{bmatrix} e_1^{meas} \\ e_2^{meas} \\ \vdots \\ e_r^{meas} \end{bmatrix} \qquad (6.11)$$

Digression: Examples of the Kalman Model

The Kalman filter is surprisingly simple to write down, but its algebraic derivation is fraught with complexity.* To fortify the reader for the task, we are going to list a number of processes that are susceptible to the formulation (6.10) and (6.11). The astonishing variety of instances where the filter has been applied is quite impressive and may make the subsequent derivations more palatable.

i. The experiment analyzed in Section 6.2 simply involved the estimation of a constant; the weight of the ring did not change between the measurements by the jewelers, and the weight was measured directly. Therefore,

$$X_{new}^{true} = \begin{bmatrix} 1 \end{bmatrix} X_{old}^{true} \quad \text{and} \quad S = \begin{bmatrix} 1 \end{bmatrix} X_{new}^{true} + e_{meas}.$$

ii. A *constant velocity* tracking process refers to a one-dimensional motion where the object's velocity v is constant between position measurements, except for random perturbations of the motion by such effects as wind, tremors, and the like. The transition between positions X over a time interval Δt is formulated as

$$X_{new}^{true} = X_{old}^{true} + \begin{bmatrix} v\Delta t \end{bmatrix} + e_{trans}, \quad S = \begin{bmatrix} 1 \end{bmatrix} X_{new}^{true} + e_{meas}.$$

The Kalman parameters are $A = [1]$, $B = [v\Delta t]$, and $D = [1]$.

iii. A *constant acceleration* tracking process requires keeping track of *both* position (X_1) and velocity (X_2). If the acceleration is denoted a, then elementary calculus tells us the details of the unperturbed transition; the noisy transition formulation is

$$X_{new}^{true} \equiv \begin{bmatrix} X_1 \\ X_2 \end{bmatrix}_{new}^{true} = \begin{bmatrix} X_1 \\ X_2 \end{bmatrix}_{old}^{true} + \begin{bmatrix} X_{2\,old}^{true}\Delta t + \frac{1}{2}a(\Delta t)^2 \\ a\Delta t \end{bmatrix} + \begin{bmatrix} e_1 \\ e_2 \end{bmatrix}_{trans}$$

$$= \underbrace{\begin{bmatrix} 1 & \Delta t \\ 0 & 1 \end{bmatrix}}_{A} \begin{bmatrix} X_1 \\ X_2 \end{bmatrix}_{old}^{true} + \underbrace{\begin{bmatrix} \frac{1}{2}a(\Delta t)^2 \\ a\Delta t \end{bmatrix}}_{B} + \underbrace{\begin{bmatrix} e_1 \\ e_2 \end{bmatrix}_{trans}}_{e_{trans}}.$$

---

* The editors forbid the use of the word "boring."

If we can only measure the position (and not the velocity), the measurement formulation becomes

$$S = \underbrace{\begin{bmatrix} 1 & 0 \end{bmatrix}}_{D} \begin{bmatrix} X_1 \\ X_2 \end{bmatrix}^{true}_{new} + e_{meas}.$$

An early application of this formulation was the tracking of spacecraft (in three dimensions, of course), with onboard inertial navigation gyros providing the value of a, and celestial or GPS readings providing the measurement S. (The gyro uncertainty was incorporated into e_{trans}.) The Apollo project incorporated the first important use of the Kalman filter.

ONLINE SOURCES

A historical account of the implementation of the Kalman filter in the Apollo is available at

McGee, L.A. and Schmidt, S.F. Discovery of the Kalman Filter as a practical tool for aerospace and industry. NASA Technical Memorandum 86847, Nov. 1985. Accessed June 24, 2016. https://ntrs.nasa.gov/archive/nasa/casi.ntrs.nasa.gov/19860003843.pdf.

iv. A frequent occurrence of a transition of the form (6.10) is due to the evolution of the variable X in accordance with some linear differential equation. The general solution of linear *first-order* differential equations of the form

$$\frac{dX}{dt} = a(t)X + g(t) \tag{6.12}$$

can be expressed (see any good differential equations textbook)*

$$X(t) \equiv e^{\int_{t_0}^{t} a(\tau)d\tau} \left[\int_{t_0}^{t} e^{-\int_{t_0}^{\tau} a(\tau)d\tau} g(\tau)d\tau + X(t_0) \right].$$

Therefore, with $t_{new} = t$ and $t_{old} = t_0$, the transition can be formulated

$$X^{true}_{new} = \underbrace{e^{\int_{t_0}^{t} a(\tau)d\tau}}_{A} X^{true}_{old} + \underbrace{e^{\int_{t_0}^{t} a(\tau)d\tau} \int_{t_0}^{t} e^{-\int_{t_0}^{\tau} a(\tau)d\tau} g(\tau)d\tau}_{B}.$$

---

* The best is *Fundamentals of Differential Equations*, by Nagle, R.K., Saff, E.B., and Snider, A.D. 2016. 7th edn. Boston, MA: Pearson.

If $g(t)$ has deterministic and random components, $g(t) = g_{det}(t) + g_{noise}(t)$, the Kalman formulation would look like

$$X_{new}^{true} = \underbrace{e^{\int_{t_0}^{t} a(\tau)d\tau}}_{A} X_{old}^{true} + \underbrace{e^{\int_{t_0}^{t} a(\tau)d\tau} \int_{t_0}^{t} e^{-\int_{t_0}^{\tau} a(\tau)d\tau} g_{det}(\tau)d\tau}_{B} + \underbrace{e^{\int_{t_0}^{t} a(\tau)d\tau} \int_{t_0}^{t} e^{-\int_{t_0}^{\tau} a(\tau)d\tau} g_{noise}(\tau)d\tau}_{e_{trans}}$$

For example, consider the RC circuit in Figure 6.1.

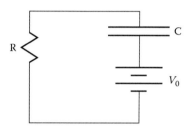

FIGURE 6.1   RC circuit.

The voltage V across the capacitor is governed by the differential equation (Kirchhoff's current law):

$$C\frac{dV}{dt} = -\frac{V + V_0}{R} \quad \text{or} \quad \frac{dV}{dt} = -\frac{V}{RC} - \frac{V_0}{RC},$$

whose solution is

$$V(t) = \underbrace{e^{-t/RC}V(0)}_{A} + \underbrace{V_0\left[e^{-t/RC} - 1\right]}_{B}.$$

The generic differential Equation 6.12 describes RL circuits, radioactivity, heat exchange, and (in a stochastic setting) financial spot prices.*

v. An RLC circuit or a forced mass-spring oscillator evolves according to a *second*-order linear differential equation like

$$\frac{d^2X}{dt^2} = a(t)\frac{dX}{dt} + b(t)X + g(t) \tag{6.13}$$

---

* Arnold, T., Bertus, M., and Godbey, J. 2008. A simplified approach to understanding the Kalman Filter technique. *The Engineering Economist* 53: 140–155.

and its general solution takes the form

$$X(t) = c_1 X_{hom}^{(1)}(t) + c_2 X_{hom}^{(2)}(t) + X_{part}(t),$$ (6.14)

where

$X_{hom}^{(1)}(t)$ and $X_{hom}^{(2)}(t)$ are independent solutions of the associated homogeneous equation

$X_{part}(t)$ is a particular solution

c_1 and c_2 are arbitrary constants*

Equation 6.13 carries *two* initial conditions with it, $X(0)$ and $X'(0)$, and the solution (6.14) can be rearranged into the generic form

$$X(t) = \tilde{X}_1(t) X(0) + \tilde{X}_2(t) X'(0) + \tilde{X}_3(t).$$ (6.15)

To mold this formulation into the Kalman format, we append the differentiated form of (6.15),

$$X'(t) = \tilde{X}_1'(t) X(0) + \tilde{X}_2'(t) X'(0) + \tilde{X}_3'(t),$$ (6.16)

and express (6.15) and (6.16) as a matrix equation for the vector $[X(t)\ X'(t)]^T$:

$$\begin{bmatrix} X(t) \\ X'(t) \end{bmatrix} = \begin{bmatrix} \tilde{X}_1(t) & \tilde{X}_2(t) \\ \tilde{X}_1'(t) & \tilde{X}_2'(t) \end{bmatrix} \begin{bmatrix} X(0) \\ X'(0) \end{bmatrix} + \begin{bmatrix} \tilde{X}_3(t) \\ \tilde{X}_3'(t) \end{bmatrix}.$$ (6.17)

With noise, (6.17) takes the Kalman form $X_{new}^{true} = AX_{old}^{true} + B + e_{trans}$.

vi. The solution to a linear *system* of first-order equations in matrix form

$$\frac{d}{dt}[X] = M[X] + [g(t)]$$

can be expressed using the matrix exponential,[†] if the coefficient matrix M is constant:

$$X(t) = \underbrace{[e^{Mt}][e^{-Mt_0}]}_{A}[X(t_0)] + \underbrace{[e^{Mt}]\int_{t_0}^{t}[e^{-M\tau}]g(\tau)d\tau}_{B}.$$

Thus, breaking $g(t)$ into deterministic and noise components, we again obtain the Kalman formulation.

---

* See *Fundamentals of Differential Equations*, op. cit.
[†] See *Fundamentals of Differential Equations*, op cit. or *Fundamentals of Matrix Analysis*, op cit.

vii. The ARMA processes of Section 4.3 fit readily into the Kalman scheme. For ARMA(3,0), we have

$$X(n) = a(1)X(n-1) + a(2)X(n-2) + a(3)X(n-3) + b(0)V(n),$$

which we recast as

$$\underbrace{\begin{bmatrix} X(n) \\ X(n-1) \\ X(n-2) \end{bmatrix}}_{X_{new}^{true}} = \underbrace{\begin{bmatrix} a(1) & a(2) & a(3) \\ 1 & 0 & 0 \\ 0 & 1 & 0 \end{bmatrix}}_{A} \underbrace{\begin{bmatrix} X(n-1) \\ X(n-2) \\ X(n-3) \end{bmatrix}}_{X_{old}^{true}} + \underbrace{\begin{bmatrix} b(0)V(n) \\ 0 \\ 0 \end{bmatrix}}_{e_{trans}}.$$

All of these examples have involved linear transitions—described by matrices. Although we will not go into them in this text, Kalman and others have extended the technique to nonlinear transitions. So, the Kalman filter is certainly a robust tool. The effort you are about to spend in deriving its equations will pay off.

6.3   SCALAR KALMAN FILTER WITH NOISELESS TRANSITION

We assume that we have noisy unbiased estimates (X_{old}, X_{new}) of the true values

$$X_{old} - X_{old}^{true} = e_{old}, \quad X_{new} - X_{new}^{true} = e_{new},$$

with zero-mean independent errors as before. Now the two competing estimates for X_{new}^{true} are (1) X_{new} and (2) the old estimate X_{old}, updated by the transition equation $AX_{old} + B$. So, the Kalman estimate will be a combination of these:

$$X_{Kalman} = KX_{new} + L\left[AX_{old} + B\right]. \tag{6.18}$$

For (6.18) to be unbiased, the expected values must satisfy

$$E\{X_{Kalman}\} = E\{KX_{new} + L[AX_{old} + B]\} = KX_{true}^{new} + L\left[AX_{old}^{true} + B\right] = (K + L)X_{true}^{new},$$

so $L = 1 - K$ as before

$$X_{Kalman} = KX_{new} + (1 - K)\left[AX_{old} + B\right]. \tag{6.19}$$

To express the error, we subtract the tautology

$$X_{new}^{true} = KX_{new}^{true} + (1 - K)X_{new}^{true} = KX_{new}^{true} + (1 - K)\left[AX_{old}^{true} + B\right]$$

from (6.19) and find

$$X_{Kalman} - X_{new}^{true} = K\left[X_{new} - X_{new}^{true}\right] + (1-K)A\left(X_{old} - X_{old}^{true}\right) = Ke_{new} + (1-K)Ae_{old}.$$

This is identical to the error expression (6.5) in Section 6.1, except for the extra factor A that accompanies e_{old}. Therefore, to express the Kalman gain and mean squared error, we simply need to modify (6.7) and (6.8) of Section 6.1 accordingly:

$$K_{Kalman} = \frac{A^2\sigma_{old}^2}{\sigma_{new}^2 + A^2\sigma_{old}^2}, \tag{6.20}$$

$$\sigma_{Kalman}^2 = \frac{A^2\sigma_{old}^2}{\sigma_{new}^2 + A^2\sigma_{old}^2}\sigma_{new}^2 = K_{Kalman}\sigma_{new}^2. \tag{6.21}$$

6.4   SCALAR KALMAN FILTER WITH NOISY TRANSITION

The next generalization of the Kalman filter allows noise in the transition process:

$$X_{new}^{true} = AX_{old}^{true} + B + e_{trans}. \tag{6.22}$$

For example, the RC circuit of Section 6.2 was described by the transition equation

$$V_{new} = AV_{old} + B \quad \text{with } A = e^{-t/RC} \text{ and } B = V_0\left[e^{-t/RC} - 1\right],$$

and if the determination of the source voltage V_0 was subject to error, then $V_0 = V_0^{true} + e_0$ and

$$V_{new} = AV_{old} + B + e_{transition} \quad \text{with } A = e^{-t/RC}, B = V_0^{true}\left[e^{-t/RC} - 1\right],$$

$$\text{and } e_{trans} = e_0\left[e^{-t/RC} - 1\right].$$

The noise e_{trans} is assumed to be zero-mean and independent of the errors in the measurements. Note that this makes X_{new}^{true} a *random* variable, and it has a mean

$$\overline{X_{new}^{true}} = E\left\{X_{new}^{true}\right\} = E\left\{AX_{old}^{true} + B + e_{trans}\right\} = AX_{old}^{true} + B + E\left\{e_{trans}\right\} = AX_{old}^{true} + B.$$

In Figure 6.2, we have attached superscripts to the variables to distinguish between measured, true, and mean values. All the error terms have mean zero and are independent.

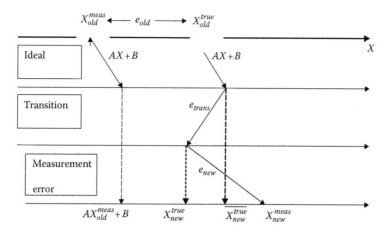

FIGURE 6.2   Kalman filter with noisy transition (read down from top.).

We have to construct the Kalman estimator of (the random variable) X_{new}^{true} using the available data X_{old}^{meas} and X_{new}^{meas}. We choose, as before,

$$X_{Kalman} = KX_{new}^{meas} + L\left[AX_{old}^{meas} + B \right]. \qquad (6.23)$$

Now, we cannot enforce the condition $E\{X_{Kalman}\} = X_{new}^{true}$ because X_{new}^{true} is random. Therefore, we impose the bias condition $E\{X_{Kalman}\} = E\{X_{new}^{true}\}$, and our analysis is slightly modified; however, since the error terms all have zero mean, we can insert them judiciously to write

$$E\{X_{Kalman}\} = E\left\{ KX_{new}^{meas} + L\left[AX_{old}^{meas} + B \right] \right\}$$

$$= E\left\{ K\left(X_{new}^{true} + e_{new} \right) \right\} + E\left\{ L\left[A\left(X_{old}^{true} + e_{old} \right) + B \right] \right\}.$$

To manipulate this expression into a form embracing the *noisy* transition Equation 6.22, we add and subtract e_{trans} and rearrange to derive

$$E\{X_{Kalman}\} = KE\{X_{new}^{true}\} + K \cdot 0 + LE\left\{ \left[A\left(X_{old}^{true} \right) + B + e_{trans} - e_{trans} + Ae_{old} \right] \right\}$$

$$= KE\{X_{new}^{true}\} + LE\{X_{new}^{true}\} - L \cdot 0 + LA \cdot 0$$

$$= (K + L)E\{X_{new}^{true}\},$$

and we are led to take $L = 1 - K$ again

$$X_{Kalman} = KX_{new}^{meas} + (1 - K)\left[AX_{old}^{meas} + B \right]. \qquad (6.24)$$

To analyze the error in the estimator (6.24), we subtract the tautology

$$X_{new}^{true} = KX_{new}^{true} + (1-K)X_{new}^{true} = KX_{new}^{true} + (1-K)\left[AX_{old}^{true} + B + e_{trans}\right]$$

and derive

$$e_{Kal} = X_{Kalman} - X_{new}^{true} = K\left[X_{new}^{meas} - X_{new}^{true}\right] + (1-K)\left[A\left(X_{old}^{meas} - X_{old}^{true}\right) - e_{trans}\right]$$

$$= Ke_{new} + (1-K)Ae_{old} - (1-K)e_{trans}. \tag{6.25}$$

The mean squared error in the Kalman estimator is then

$$E\left\{\left[X_{Kalman} - X_{new}^{true}\right]^2\right\} = E\left\{\left[Ke_{new} + (1-K)Ae_{old} - (1-K)e_{trans}\right]^2\right\}$$

$$= K^2 E\left\{e_{new}^2\right\} + (1-K)^2 A^2 E\left\{e_{old}^2\right\} + (1-K)^2 E\left\{e_{trans}^2\right\}$$

$$\left(\text{since the errors are uncorrelated}\right)$$

$$= K^2\sigma_{new}^2 + (1-K)^2\left[A^2\sigma_{old}^2 + \sigma_{trans}^2\right]. \tag{6.26}$$

This expression is identical to (6.6) in Section 6.1 with the replacement of σ_{old}^2 by $A^2\sigma_{old}^2 + \sigma_{trans}^2$ so, making this substitution in (6.7) and (6.8), we find the Kalman gain to be

$$K_{Kalman} = \frac{A^2\sigma_{old}^2 + \sigma_{trans}^2}{\sigma_{new}^2 + A^2\sigma_{old}^2 + \sigma_{trans}^2} \tag{6.27}$$

and the LMSE to be

$$\sigma_{Kalman}^2 = \frac{A^2\sigma_{old}^2 + \sigma_{trans}^2}{\sigma_{new}^2 + A^2\sigma_{old}^2 + \sigma_{trans}^2}\sigma_{new}^2$$

$$= K_{Kalman}\sigma_{new}^2. \tag{6.28}$$

6.5   ITERATION OF THE SCALAR KALMAN FILTER

In most of the applications of Kalman filtering listed in Section 6.2, the system under study is dynamic, undergoing a *series* of noisy transitions, and we need to update the Kalman estimate for X after each one. This seems straightforward enough; we line up two copies of Figure 6.2 and associate the previous "new" values with the current "old" values; see Figure 6.3.

But there are two modifications that are necessary:

i. As we have noted, the previous X_{new}^{true}, which is destined to be the current X_{old}^{true}, is random.

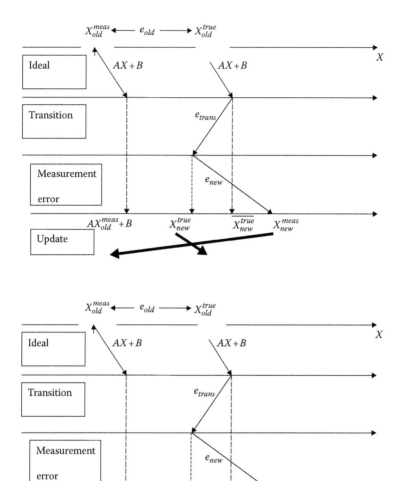

FIGURE 6.3   Consecutive Kalman filtering.

ii. Instead of using the previous *measured value* in the transition equation, we
should use the previous X_{Kalman} (since it has lower mean square error),

$$X_{Kalman} = K X_{new}^{meas} + L\left[A X_{old}^{Kal} + B \right].$$

Thus, the situation is more honestly described by Figure 6.4.

Fortunately, this doesn't change the equation for the gain or the LMSE; the bias
condition now looks like

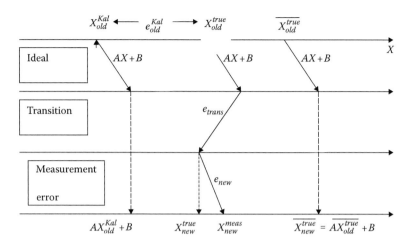

FIGURE 6.4 Consecutive Kalman filtering (corrected).

$$E\left\{X_{Kalman}\right\} = E\left\{KX_{new}^{meas} + L\left[AX_{old}^{Kal} + B\right]\right\}$$

$$= E\left\{K\left(X_{new}^{true} + e_{new}\right)\right\} + E\left\{L\left[A\left(X_{old}^{true} + e_{old}^{Kal}\right) + B\right]\right\}$$

$$= KE\left\{X_{new}^{true}\right\} + K\cdot 0 + LE\left\{\left[A\left(X_{old}^{true}\right) + B + e_{trans} - e_{trans} + Ae_{old}^{Kal}\right]\right\}$$

$$= KE\left\{X_{new}^{true}\right\} + LE\left\{X_{new}^{true}\right\} - L\cdot 0 + LA\cdot 0$$

$$= \left(K + L\right)E\left\{X_{new}^{true}\right\},$$

requiring $L = 1 - K$ yet again

$$X_{Kalman} = KX_{new}^{meas} + \left(1 - K\right)\left[AX_{old}^{Kal} + B\right].$$

Subtracting the tautology

$$X_{new}^{true} = KX_{new}^{true} + \left(1 - K\right)X_{new}^{true} = KX_{new}^{true} + \left(1 - K\right)\left[AX_{old}^{true} + B + e_{trans}\right],$$

we recover an error formula similar to that of the preceding section

$$e_{Kal} = X_{Kalman} - X_{new}^{true} = K\left[X_{new}^{meas} - X_{new}^{true}\right] + \left(1 - K\right)\left[A\left(X_{old}^{Kal} - X_{old}^{true}\right) - e_{trans}\right]$$

$$= Ke_{new} + \left(1 - K\right)Ae_{old}^{Kal} - \left(1 - K\right)e_{trans}. \tag{6.29}$$

The only difference between (6.29) and (6.25) of Section 6.4 is the replacement of e_{old} by e_{old}^{Kal}. Therefore, the modifications to the gain and LMSE formulas (6.27) and (6.28) are

$$K_{Kalman} = \frac{A^2 \left(\sigma_{old}^{Kal}\right)^2 + \sigma_{trans}^2}{\sigma_{new}^2 + A^2 \left(\sigma_{old}^{Kal}\right)^2 + \sigma_{trans}^2} \qquad (6.30)$$

and

$$\left(\sigma_{new}^{Kal}\right)^2 = \frac{A^2 \left(\sigma_{old}^{Kal}\right)^2 + \sigma_{trans}^2}{\sigma_{new}^2 + A^2 \left(\sigma_{old}^{Kal}\right)^2 + \sigma_{trans}^2} \, \sigma_{new}^2 = K_{Kalman}\sigma_{new}^2. \qquad (6.31)$$

6.6 MATRIX FORMULATION FOR THE KALMAN FILTER

We culminate our exposition of the Kalman filter by permitting X, B, and e_{trans} to be p-by-1 vectors and A to be a p-by-p matrix, satisfying the transition equation

$$X_{new}^{true} = A X_{old}^{true} + B + e_{trans}. \qquad (6.32)$$

The (noisy) measurements of X_{new}^{true} result in an r-by-1 vector S of the form

$$S = D X_{new}^{true} + e_{meas} \qquad (6.33)$$

where D is r-by-p and e_{meas} is r-by-1.

Remember: If D is the identity matrix, we are measuring each component separately. If $r > p$, the measurements are redundant—or, at least, they would be if there were no noise. If $r < p$, the number of measurements is insufficient to determine X completely (even in the absence of noise). All of these situations can occur in practice.

We are given the following:

i. An estimate of the previous value of the vector, X_{old}^{est}, obtained either through measurement X_{old}^{meas} or as the result of a prior Kalman estimate X_{old}^{Kal}
ii. The matrices A and B describing the transition model (6.32)
iii. The result S of the current measurement (6.33)

We propose to construct the Kalman estimator of X_{new}^{true} as a combination of S and the updated previous estimate,

$$X_{Kalman} = K S + L \left[A X_{old}^{est} + B \right], \tag{6.34}$$

where K and L are p-by-r and p-by-p matrices, respectively—to be determined.

The error vectors in the formulas

$$X_{old}^{est} = X_{old}^{true} + e_{old}, \quad X_{new}^{true} = AX_{old}^{true} + B + e_{trans}, \quad S = DX_{new}^{true} + e_{meas} \tag{6.35}$$

are assumed to be zero-mean. Characterizing their correlations, however, is more subtle. First note that multiplication of a *column* error vector by a *row* error vector yields a matrix of all the componentwise products. For example,

$$
\left[e_{old} \right] \left[e_{meas} \right]^T \equiv
\begin{bmatrix}
e_1^{old} \\
e_2^{old} \\
\vdots \\
e_p^{old}
\end{bmatrix}
\begin{bmatrix}
e_1^{meas} & e_2^{meas} & \cdots & e_r^{meas}
\end{bmatrix}
$$

$$
=
\begin{bmatrix}
e_1^{old} e_1^{meas} & e_1^{old} e_2^{meas} & \cdots & e_1^{old} e_r^{meas} \\
e_2^{old} e_1^{meas} & e_2^{old} e_2^{meas} & \cdots & e_2^{old} e_r^{meas} \\
& \ddots & & \\
e_p^{old} e_1^{meas} & e_p^{old} e_2^{meas} & \cdots & e_p^{old} e_r^{meas}
\end{bmatrix}. \tag{6.36}
$$

We assume that all of the error components in the estimate of X_{old}^{true} are independent of all of the error components in the new measurement; hence, by (6.36),

$$E\left\{ \left[e_{old} \right] \left[e_{meas} \right]^T \right\} = 0 \quad \left(\text{the zero matrix} \right). \tag{6.37}$$

Similarly, the other error processes are independent of each other:

$$E\left\{ \left[e_{old} \right] \left[e_{trans} \right]^T \right\} = 0, \quad E\left\{ \left[e_{trans} \right] \left[e_{meas} \right]^T \right\} = 0. \tag{6.38}$$

However, the various components of the errors in *one particular* process may be correlated. Therefore, we have the *error autocorrelation matrices*

$$
E\left\{\left[e_{old}\right]\left[e_{old}\right]^{T}\right\} = E\left\{\begin{bmatrix} e_1^{old}\,e_1^{old} & e_1^{old}\,e_2^{old} & \cdots & e_1^{old}\,e_p^{old} \\ e_2^{old}\,e_1^{old} & e_2^{old}\,e_2^{old} & \cdots & e_2^{old}\,e_p^{old} \\ & & \ddots & \\ e_p^{old}\,e_1^{old} & e_p^{old}\,e_2^{old} & \cdots & e_p^{old}\,e_p^{old} \end{bmatrix}\right\}
$$

$$
= \begin{bmatrix} \sigma_1^{2,old} & \overline{e_1^{old}\,e_2^{old}} & \cdots & \overline{e_1^{old}\,e_p^{old}} \\ \overline{e_2^{old}\,e_1^{old}} & \sigma_2^{2,old} & \cdots & \overline{e_2^{old}\,e_p^{old}} \\ & & \ddots & \\ \overline{e_p^{old}\,e_1^{old}} & \overline{e_p^{old}\,e_2^{old}} & \cdots & \sigma_p^{2,old} \end{bmatrix} \equiv Q_{old}
$$

$$
E\left\{\left[e_{trans}\right]\left[e_{trans}\right]^{T}\right\} = \begin{bmatrix} \sigma_1^{2,trans} & \overline{e_1^{trans}\,e_2^{trans}} & \cdots & \overline{e_1^{trans}\,e_p^{trans}} \\ \overline{e_2^{trans}\,e_1^{trans}} & \sigma_2^{2,trans} & \cdots & \overline{e_2^{trans}\,e_p^{trans}} \\ & & \ddots & \\ \overline{e_p^{trans}\,e_1^{trans}} & \overline{e_p^{trans}\,e_2^{trans}} & \cdots & \sigma_p^{2,trans} \end{bmatrix} \equiv Q_{trans}
$$

$$
E\left\{\left[e_{meas}\right]\left[e_{meas}\right]^{T}\right\} = \begin{bmatrix} \sigma_1^{2,meas} & \overline{e_1^{meas}\,e_2^{meas}} & \cdots & \overline{e_1^{meas}\,e_r^{meas}} \\ \overline{e_2^{meas}\,e_1^{meas}} & \sigma_2^{2,meas} & \cdots & \overline{e_2^{meas}\,e_r^{meas}} \\ & & \ddots & \\ \overline{e_r^{meas}\,e_1^{meas}} & \overline{e_r^{meas}\,e_2^{meas}} & \cdots & \sigma_r^{2,meas} \end{bmatrix} \equiv Q_{meas},
$$

and we shall *not* assume that the off-diagonal elements of the error autocorrelation matrices Q_{old}, Q_{trans}, and Q_{meas} are zero. (Thus, we allow for the possibility that the various measurements in S, or the various transition error components, might interfere with each other.) The issue of how to come up with these autocorrelation matrices depends on the particular situation. The case studies contain examples.

We regard the total mean squared error in e_{old} to be the sum of squares of the standard deviations of the components of e_{old}, and similarly for e_{trans} and e_{meas}. These total errors are given by the sums of the diagonal elements of the corresponding autocorrelation matrices Q_{old}, Q_{trans}, and Q_{meas}. Such a sum is called the *trace* of the matrix.

The matrices K and L in the Kalman estimator (6.34) have to be chosen so that the estimate is unbiased and has LMSE. The bias condition is

$$
\begin{aligned}
E\left\{X_{Kalman}\right\} &= E\left\{KS + L\left[AX_{old}^{est} + B\right]\right\} \\
&= E\left\{K\left(DX_{new}^{true} + e_{meas}\right)\right\} + E\left\{L\left[A\left(X_{old}^{true} + e_{old}\right) + B\right]\right\} \\
&= KDE\left\{X_{new}^{true}\right\} + K\cdot 0 + LE\left\{\left[A\left(X_{old}^{true}\right) + B + e_{trans} - e_{trans} + A\,e_{old}\right]\right\}, \\
&= KDE\left\{X_{new}^{true}\right\} + LE\left\{X_{new}^{true}\right\} - L\cdot 0 + LA\cdot 0 \\
&= \left(KD + L\right)E\left\{X_{new}^{true}\right\}, \tag{6.39}
\end{aligned}
$$

so we require that $KD + L = I$ (the identity matrix), or $L = I - KD$. Therefore, the estimator can be rewritten in the form

$$X_{Kalman} = \left[AX_{old}^{est} + B \right] + K \left[S - D \left(A X_{old}^{est} + B \right) \right],$$ (6.40)

with K interpreted as the *Kalman gain matrix*.

To analyze the error in the Kalman predictor, we subtract the tautology

$$X_{new}^{true} = X_{new}^{true} + K \left(DX_{new}^{true} - DX_{new}^{true} \right)$$

$$= \left[AX_{old}^{true} + B + e_{trans} \right] + K \left(DX_{new}^{true} - D \left[A X_{old}^{true} + B + e_{trans} \right] \right)$$

from (6.40) and bulldoze through the algebra to derive

$$e_{Kal} = X_{Kalman} - X_{new}^{true} = Ae_{old} - e_{trans} + K \left[e_{meas} - D \left(Ae_{old} - e_{trans} \right) \right].$$ (6.41)

Since we wish to minimize the total mean squared error in (6.41), we need to choose the Kalman gain matrix to minimize the trace of

$$Q_{Kal} = E \left\{ e_{Kal} e_{Kal}^T \right\}$$

$$= E \left\{ \left(Ae_{old} - e_{trans} + K \left[e_{meas} - D \left(Ae_{old} - e_{trans} \right) \right] \right) \right.$$

$$\times \left(Ae_{old} - e_{trans} + K \left[e_{meas} - D \left(Ae_{old} - e_{trans} \right) \right] \right)^T \right\}$$

$$= E \left\{ \left(Ae_{old} - e_{trans} + K \left[e_{meas} - D \left(Ae_{old} - e_{trans} \right) \right] \right) \right.$$

$$\times \left(e_{old}^T A^T - e_{trans}^T + \left[e_{meas}^T - \left(e_{old}^T A^T - e_{trans}^T \right) D^T \right] K^T \right) \right\}.$$ (6.42)

Because the crosscorrelation matrices like $E \left\{ e_{old} e_{trans}^T \right\}$ are zero, (6.42) simplifies to

$$Q_{Kal} = AQ_{old} A^T + Q_{trans} - \left(AQ_{old} A^T + Q_{trans} \right) D^T K^T - KD \left(AQ_{old} A^T + Q_{trans} \right)$$

$$+ K \left[Q_{meas} + D \left(AQ_{old} A^T + Q_{trans} \right) D^T \right] K^T$$ (6.43)

At the minimum point, the partial derivative of tr$[Q_{Kal}]$ with respect to each component of the matrix K will be zero. This is straightforward enough to work out for (6.43) term by term, but it is enlightening to retain the matrix formulation. The first two terms of (6.43) are independent of K, so they are not a problem. The final three terms have the form tr$[GK^T]$, tr$[KG]$, and tr$[KGK^T]$, where G is a generic constant matrix of the appropriate dimensions; so we have to figure out how to differentiate such terms.

First, we focus on the term K_{ij} in tr$[KG]$:

$$\mathrm{tr}\{KG\} = \mathrm{tr}\left\{\begin{bmatrix} & \cdots & (col\#\,j) & \cdots \\ (row\#\,i) & & K_{ij} & \cdots \\ & \cdots & & \end{bmatrix}\begin{bmatrix} \cdots & \cdots & \cdots \\ \cdots & G & \cdots \\ \cdots & \cdots & \cdots \end{bmatrix}\right\}.$$

The partial derivative with respect to K_{ij} (holding all other elements of K fixed) will equal

$$\frac{\partial \mathrm{tr}\{KG\}}{\partial K_{ij}} = \mathrm{tr}\left\{\begin{bmatrix} 0 & \cdots & (col\#\,j) & \cdots & 0 \\ & & \vdots & & \\ (row\#\,i) & \cdots & 1 & \cdots & \cdots \\ & & \vdots & & \\ 0 & \cdots & \cdots & \cdots & 0 \end{bmatrix}\begin{bmatrix} & & \\ & G & \\ & & \end{bmatrix}\right\}$$

$$= \mathrm{tr}\left\{\begin{bmatrix} 0 & \cdots & 0 & \cdots & 0 \\ & & \vdots & & \\ \cdots & \cdots & G\text{'s } jth\ row & \cdots & \cdots \\ & & \vdots & & \\ 0 & \cdots & 0 & \cdots & 0 \end{bmatrix} \leftarrow row\ position\#\,i\right\}$$

$$= G_{ji}. \tag{6.44}$$

Thus, if we define $\dfrac{\partial \mathrm{tr}\{KG\}}{\partial K}$ to be a matrix whose i,jth entry is $\partial \mathrm{tr}\{KG\}/\partial K_{ij}$, (6.44) states that

$$\frac{\partial \mathrm{tr}\{KG\}}{\partial K} = G^T. \tag{6.45}$$

Now, since the diagonals of a matrix remain unmoved when one takes the transpose, tr$[GK^T]$ = tr$[(GK^T)^T]$ = tr$[KG^T]$ and it follows from (6.45) that

$$\frac{\partial \mathrm{tr}\{GK^T\}}{\partial K} = \frac{\partial \mathrm{tr}\{KG^T\}}{\partial K} = \left[G^T\right]^T = G. \tag{6.46}$$

For the quadratic form tr$[KGK^T]$, we use the product rule. That is, we regard the "second" K as constant and differentiate with respect to the "first" K, and then reverse the roles, and add.

$$\frac{\partial \text{tr}\left\{KGK^T\right\}}{\partial K_{ij}} = \left.\frac{\partial \text{tr}\left\{KG\tilde{K}^T\right\}}{\partial K_{ij}}\right|_{\tilde{K}\text{ fixed},=K} + \left.\frac{\partial \text{tr}\left\{\tilde{K}GK^T\right\}}{\partial K_{ij}}\right|_{\tilde{K}\text{ fixed},=K}.$$

Thus, applying (6.45) and (6.46),

$$\frac{\partial \text{tr}\left\{KGK^T\right\}}{\partial K} = \left[GK^T\right]^T + KG = K\left(G + G^T\right). \tag{6.47}$$

Consequently, the sum of the mean squared errors of the Kalman estimator is minimized when

$$\frac{\partial \, \text{tr}\left[Q_{Kal}\right]}{\partial K} = 0$$

$$= \frac{\partial \, \text{tr}\left[AQ_{old}A^T + Q_{trans}\right]}{\partial K} - \frac{\partial \, \text{tr}\left[\left(AQ_{old}A^T + Q_{trans}\right)D^T K^T\right]}{\partial K}$$

$$- \frac{\partial \, \text{tr}\left[KD\left(AQ_{old}A^T + Q_{trans}\right)\right]}{\partial K} + \frac{\partial \, \text{tr}\left[K\left[Q_{meas} + D\left(AQ_{old}A^T + Q_{trans}\right)D^T\right]K^T\right]}{\partial K}$$

$$= (0) - \left(AQ_{old}A^T + Q_{trans}\right)D^T - \left[D\left(AQ_{old}A^T + Q_{trans}\right)\right]^T$$

$$+ K\left\{\left[Q_{meas} + D\left(AQ_{old}A^T + Q_{trans}\right)D^T\right] + \left[Q_{meas} + D\left(AQ_{old}A^T + Q_{trans}\right)D^T\right]^T\right\}$$

$$= -2\left(AQ_{old}A^T + Q_{trans}\right)D^T + 2K\left[Q_{meas} + D\left(AQ_{old}A^T + Q_{trans}\right)D^T\right],$$

and therefore the Kalman gain is given by

$$K = \left(AQ_{old}A^T + Q_{trans}\right)D^T \left[Q_{meas} + D\left(AQ_{old}A^T + Q_{trans}\right)D^T\right]^{-1}. \tag{6.48}$$

The autocorrelation matrix of the error in the Kalman estimator is found by substituting (6.48) into (6.43). Straightforward algebra results in the equivalent expressions (Problem 15)

$$Q_{Kal} = \left(AQ_{old}A^T + Q_{trans}\right)\left(I - D^T K^T\right) \equiv \left(I - KD\right)\left(AQ_{old}A^T + Q_{trans}\right). \tag{6.49}$$

Note that if the components of X_{new}^{true} are measured individually, so that $S = X_{new}^{true} + e_{meas}$ and $D = I$, then

$$Q_{Kal} = \left(I - K\right)\left(AQ_{old}A^T + Q_{trans}\right)$$

$$= \left(\left[Q_{meas} + AQ_{old}A^T + Q_{trans}\right]\left[Q_{meas} + AQ_{old}A^T + Q_{trans}\right]^{-1}\right.$$

$$\left. -\left[AQ_{old}A^T + Q_{trans}\right]\left[Q_{meas} + AQ_{old}A^T + Q_{trans}\right]^{-1}\right)$$

$$\times \left(AQ_{old}A^T + Q_{trans}\right)$$

$$= \left(Q_{meas}\left[Q_{meas} + AQ_{old}A^T + Q_{trans}\right]^{-1}\right)\left(AQ_{old}A^T + Q_{trans}\right)$$

$$= Q_{meas}K^T = KQ_{meas}^T = KQ_{meas}, \tag{6.50}$$

analogous to (6.31) in Section 6.5.

If the Kalman estimator is iterated then (6.49), of course, replaces Q_{old} in the formalism.

ONLINE SOURCES

Remembering Kalman, R.E. (1930–2016). University of Florida, Gainesville, FL. https://www.eng.ufl.edu/news/remembering-rudolf-e-kalman-1930-2016/. Accessed July 11, 2016. (As this book goes to press, your author received the somber news that Rudolph Kalman passed away July 2, 2016.)

Case studies of tracking using Kalman filters, some with code, can be found at Simon, D. Kalman Filtering. Embedded Systems Programming. Accessed June 24, 2016. http://academic.csuohio.edu/simond/courses/eec644/kalman.pdf.
   [constant-acceleration trajectories; contains MATLAB® code for simulations.]

Jurić, D. Object tracking: Kalman Filter with ease. Code Project. Accessed June 24, 2016. http://www.codeproject.com/Articles/865935/Object-Tracking-Kalman-Filter-with-Ease.
Levy, L.J. The Kalman Filter: Navigation's integration workhorse. The Johns Hopkins University Applied Physics Lab. https://www.cs.unc.edu/~welch/kalman/Levy1997/Levy1997_KFWorkhorse.pdf.
   [navigation in general]

Saxena, A. Kalman Filter applications. Cornell University, Ithaca, NY. Accessed June 24, 2016. https://www.cs.cornell.edu/courses/cs4758/2012sp/materials/mi63slides.pdf.
   [a sloshing-tank example]

Descriptions of the use of Kalman filters in financial modeling appear in
Javaheri, A., Lautier, D., and Galli, A. 2003. Filtering in finance. *Willmot Magazine*. Accessed June 24, 2016. http://www.cis.upenn.edu/~mkearns/finread/filtering_in_finance.pdf.
   [models of commodity prices; herein are also described two methods for extending the filter to nonlinear transition models: the extended Kalman filter and the unscented Kalman filter.]

Munroe, M., Hench, T., and Timmons, M. Sales rate and cumulative sales forecasting using Kalman Filtering techniques. Accessed June 24, 2016. https://www.cs.unc.edu/~welch/kalman/media/pdf/Munroe2009.pdf.
   [sales rate prediction]

Arnold, T., Bertus, M.J., and Godbey, J. A simplified approach to understanding the Kalman Filter technique. Taylor & Francis Online. Accessed June 24, 2016. http://www.tandfon-line.com/doi/abs/10.1080/00137910802058574?queryID=%24{resultBean.queryID}. [implements the filter in Excel for a spot/future prices case study]

EXERCISES

SECTION 6.1

1. As described in Section 6.1, two jewelers make measurements of the weight of a gold earring. The first jeweler reports a value of 50 g, with an estimated root mean squared error of 0.6 g; in other words, the estimated *squared* error is 0.36. The second measurement results in a value of 50.5 g, with an estimated root mean squared error of 0.8 g. What is the best (Kalman) estimate of the actual weight of the earring, and what is its mean squared error?

2. Suppose the measurements in Problem 1 were done in the reverse order—the 50.5 ± 0.8 was done first—then the 50.0 ± 0.6 was done second. Work out the Kalman estimate for the weight of the earring. Is your answer different from #1? Discuss.

3. The position (x-coordinate) of an object is measured. Considering the reliability of the measurement, the probability distribution function for its position is as shown in the left-hand graph of Figure 6.5. A second measurement was made, resulting in the pdf as shown on the right. Use the Kalman filter to deduce the best estimate of the position and its mean squared error.

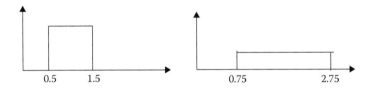

FIGURE 6.5 Measurement error distributions.

SECTION 6.4

4. Starting now, a coin is flipped every minute. If it is heads on the first flip, a marker is placed at $X = +1$; if it is tails, the marker is placed at $X = -1$. At the second flip, it is moved one unit to the right if the coin reads heads, and 1 to the left if it reads tails; but then it is shifted one-third of the way back to the origin. (So it's true position is either $X = 0$, 2/3, or $-2/3$.) The position X of the coin is measured to be +0.5, with a measuring device that is unbiased but has an error standard deviation of 1. What is the least-squares estimate for the position of the marker?

5. Suppose that in one day a radioactive substance decays to half its weight (its half-life is one day). Yesterday at noon, a sample was measured to weigh

100 ± 5 g. (Assume that 100 g was an unbiased estimate of its true weight and that 5 g was the square root of this estimate's expected squared error.)

Today at noon, its weight was measured to be 49 ± 3 g. Again, assume that this specifies an unbiased estimate and its root mean squared error.

A note has just been discovered stating that vandals had altered the sample by 1 g just before today's weighing; we don't know whether the 1 g was added or subtracted, and either possibility seems equally likely.

What is the Kalman estimate of the sample's true weight at noon today? What is the expected value of the squared error in this estimate?

6. Working backward, use the data of Problem 5 to calculate the Kalman estimate of the *original* weight of the sample, at noon yesterday.

7. A random process $X(n)$ is governed by the difference equation $X(n) = a(n-1)X(n-1) + W(n)$, where $W(n)$ is zero-mean white noise with power $\sigma_w^2 = 2$, uncorrelated with $X(k)$ for $k < n$. The coefficients $a(n)$ equal $(-1)^n/(n+1)$. Measurements of $X(n)$ are corrupted with additive zero-mean white noise $V(n)$ in accordance with $Y(n) = X(n) + V(n)$, where $\sigma_v^2 = 3$. The value $X(0)$ is *known* to be 5; no error in this estimate. What is the Kalman estimate for $X(2)$?

8. The current I in an RL circuit (Figure 6.6) is measured at two different times $t = t_0 = 0$ and $t = t_1 = 0.7$ s. The *measured* value of I at t_0 equals 5; the *measured* value of I at t_1 equals 1. But the measured values are corrupted by zero-mean noises, independent of all other parameters, having standard deviations $\sigma_0 = 1.0$ and $\sigma_1 = 1.3$. Compute the Kalman estimate of the current at time $t = t_1$, and the variance of the error in this estimate.

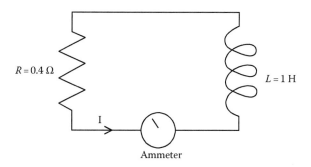

$R = 0.4\ \Omega$

$L = 1\ \text{H}$

I

Ammeter

FIGURE 6.6 RL circuit.

9. A radioactive material is known to decay with a half-life of 1 day; so if $X(n)$ is the weight of the material at noon on day n, $X(n+1) = X(n)/2$ *exactly*. The weight is measured on days 0, 1, and 2 with an *imperfect* scale that can be described statistically as follows: if the object's exact weight is μ, the weight ξ reported by the scale is $N(\mu, 0.1)$. If the measured values of weight on three consecutive days are given by $X(0) \approx 6$, $X(1) \approx 3$, and $X(2) \approx 2$, what are the iterated Kalman estimates for $X(1)$ and $X(2)$?

10. The solution to the differential equation $dy/dt = -y$ is measured at two different times t_0 and t_1, yielding values $y(t_0)$ and $y(t_1)$. But the measured values are corrupted by zero mean noise, independent of all other parameters, having standard deviation σ.

Find the formula for the Kalman estimate of $y(t_1)$ and the variance of the error in this estimate.

11. In the situation described in Problem 3, it is revealed that someone shifted the object 0.1 units between the times that the two measurements were taken. But it is not known if the object was shifted to the left or to the right; both are equally likely. *Now* use the Kalman filter to deduce the best estimate of the position, and its mean squared error.

12. Let $X(n)$ be a process that doubles on the first night, triples on the second night, and so on: $X(1) = 2X(0)$, $X(2) = 3X(1)$, ... There is no error in this process; it is exact. The values of $X(0)$, $X(1)$, and $X(2)$ are *measured*, but the measurement errors have different distributions according to Figure 6.7. The measured value of $X(0)$ equals 1, that of $X(1)$ equals 1.9, and that of $X(2)$ equals 5.8. Construct the Kalman estimate for $X(2)$.

pdf for error in first measurement   pdf for error in second and third measurement

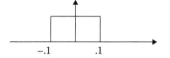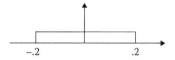

FIGURE 6.7   Measurement error distributions.

13. Reconsider the measurement problem described at the beginning of Section 6.1, but suppose the two measurement errors are not independent, having a correlation coefficient ρ. What is the optimal estimate for the weight?

14. At time $t = 0$, the switch on the RC circuit in Figure 6.8 is closed and the initial voltage on the capacitor, V_0, is drained through the resistor according to classical circuit theory. However, the measurement is corrupted by an independent noise voltage with mean zero and standard deviation σ_{meas}. Additionally, the resistor carries an independent thermal noise voltage $V_{thermal}$, described by Johnson and Nyquist (Section 2.1) as zero-mean white noise with standard

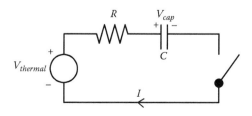

FIGURE 6.8   Noisy RC circuit.

deviation σ_{therm}. If voltage measurements are made at the times 0 and t, carry out the characterization of the Kalman filter that uses these two data to estimate the *initial* voltage on the capacitor.

SECTION 6.6

15. Derive Equation 6.49.

16. (*Contributed by the author's nephew*) S'up? This is the last problem in the book(!), so we're gonna shoot the moon. Literally. Rocket ship. Balls to the wall.

   (*Note from the author [Uncle Arthur]*: My brother's son's seemingly offensive slang is actually quite benign; "balls" in this context are the knobs atop a fighter plane's throttle control. Pushing the throttle all the way forward, to the wall of the cockpit, is to apply full throttle. Could have fooled me.)

   (*Nephew continues*) Full throttle, constant thrust J, straight up. Flat earth approximation; gravity g is constant. Force equals thrust minus gravity equals mass times acceleration.

   Hold up, hold up. The thrust isn't really constant. *Thrust jets are purchased by the space agency from the lowest bidder, dude*! Let's say it's constant for short intervals of time Δt, but the constant differs from J by a random error each time interval (independent identically distributed, zero mean).

   You know how to update the position and velocity over an interval where the acceleration is constant. So formulate the update and sort into a known part and a random part.

   Now build a Kalman filter. Be careful with the autocorrelation matrix of the transition error. It's not diagonal. Work out the case when the thrust error is white noise.

   Measurements? *We don't need no stinking measurements*. What do you think of that?

   (*Hint from author Arthur*: How would you interpret a measurement whose standard deviation was infinite?)

   How is this any better than just ignoring the thrust gremlins? (*A. A. again*: regard [6.48] and [6.49]) in the limit where Q_{meas} dominates everything else.)

Index

193